Dairy Farming: Animal Husbandry and Welfare

Published by Syrawood Publishing House,
750 Third Avenue, 9th Floor,
New York, NY 10017, USA
www.syrawoodpublishinghouse.com

Dairy Farming: Animal Husbandry and Welfare
Edited by Christian Snider

© 2016 Syrawood Publishing House

International Standard Book Number: 978-1-68286-041-0 (Hardback)

Contents

Preface

This book has been a concerted effort by a group of academicians, researchers and scientists, who have contributed their research works for the realization of the book. This book has materialized in the wake of emerging advancements and innovations in this field. Therefore, the need of the hour was to compile all the required researches and disseminate the knowledge to a broad spectrum of people comprising of students, researchers and specialists of the field.

Dairy farming, as a sub-discipline of agriculture has gained global significance. It has become one of the important economic activities in many countries around the world. This book on dairy farming and management discusses in detail various practices and modern techniques such as animal physiology, feed management, animal morphology, etc. This book provides significant information of this discipline to help develop a good understanding of dairy farming and animal welfare. Comprising state-of-the-art inputs by acclaimed experts of this field, this book targets students and professionals. This book aims to serve as a resource guide and contribute to the growth of the discipline.

At the end of the preface, I would like to thank the authors for their brilliant chapters and the publisher for guiding us all-through the making of the book till its final stage. Also, I would like to thank my family for providing the support and encouragement throughout my academic career and research projects.

Editor

Resource-Use-Efficiency among Smallholder Groundnut Farmers in Northern Region, Ghana

Gideon Danso-Abbeam[1*], Abubakari M. Dahamani[1] and Gbanha A-S Bawa[1]

[1]Department of Agricultural and Resource Economics, University for Development Studies, P.O.Box TL 1882, Tamale, Ghana.

Authors' contributions

This study was carried out in collaboration between all authors. Author GDA designed the study, performed the statistical analysis, and wrote the first draft of the manuscript. Author AMD and GASB were in charge of collection and management of the data. Author AMD managed the literature search whilst GASB streamlined the analysis of the study. All authors read and approved the final manuscript.

Editor(s):
(1) Chrysanthi Charatsari, Department of Agricultural Economics, School of Agriculture, Aristotle University of Thessaloniki, Greece.
(2) Anonymous.
Reviewers:
(1) Balde Boubacar Siddighi, The United Graduate School of Agricultural Sciences, Tottori University. Bioproduction Science, Managerial Economics. Tottori University, Japan.
(2) Nguyen Khac Minh, Water Resources University, Vietnam.

ABSTRACT

The study analyses the efficiency of resource use by collecting cross-sectional data from 120 groundnut farmers in the Tolon district of the Northern region, Ghana, during 2013 major cropping season. It focuses on identifying the determinants of groundnut output growth, measuring the technical efficiency level of the farmers as well as how efficient farmers are with respect to the allocation of their inputs. The stochastic frontier analysis *(SFA)* was employed to examine the determinants of output and measure the technical efficiency level of farmers while the marginal value product marginal factor cost *(MVP-MFC)* approach was used to ascertain whether farmers are efficiently allocating their resources or not. The results from the stochastic frontier analysis indicated that labour and quantity of seeds exerted significant and positive effects on groundnut output whilst the area of land allocated to groundnut cultivation had negative and significant effect on groundnut output. Groundnut farmers in the study area had a mean technical efficiency score of

---

Corresponding author: E-mail: nanayawdansoabbeam@gmail.com;

about 84% indicating an output loss of 16% due to inefficiency. Various sources of efficiency include; education, farming experience, household size, membership of farmer-base-organization and farmers contact with extension personnel. Allocatively, farmers were over-utilizing labour and seeds sown while under-utilizing quantities of herbicides. The study therefore recommends that an effective farm level training programmes for rural farmers through an effective extension services could increase farmer's efficiency level and hence increase their profit level.

Keywords: Stochastic frontier analysis; marginal value product marginal factor cost; technical efficiency; resource-use-efficiency; groundnut farmers; Tolon district; Ghana.

1. INTRODUCTION

According to Government of Ghana (GoG) report [1], agriculture is a vital development tool employing about 70 percent in formal and informal sector of the economy. It has accounted for an average of about 30 percent of Gross Domestic Product (GDP) in the past decade and contributed to about 60 percent of export earnings. The main driving force behind this immense contribution of agriculture is the crop sector (including cocoa) accounting for about two-thirds of the agricultural sector. Bresinger et al. [2] reported that staple crops such as maize, yam, plantain, cassava, rice, sorghum, groundnuts and oilseeds dominate the agricultural crop sector with other crops such as vegetables and fruits contributing moderately to the overall crop sector growth.

Despite such significant roles, crop productivity in Ghana has remained low. Growth in agricultural output over the years has come as a result of increase in land under cultivation rather than improvement in yields. Studies in Ghana and other parts of Africa have shown that most farm yields are lower than their achievable yields with sufficient farm management practices. For instance, in Kenya, evidence shows yield gaps of 67% for maize, 78% for groundnuts and 67% for sorghum [3]. Moreover, the average yield of rice, groundnut, yam and maize in Ghana are estimated at 2.4 mt/ha, 1.5mt/ha, 15.3 mt/ha and 1.7mt/ha respectively whereas the potential have been estimated to be 6.5 mt/ha, 2.5 mt/ha, 49 mt/ha and 6 mt/ha respectively [4]. Biophysical factors, input utilizations, socio-economic factors, management practices, climatic conditions, policy and institutional constraints as well as inadequate efforts to transfer technological knowledge and market information asymmetry were identified as some of the major reasons ascribed to these yield gaps. According to Rockstrom et al. [5], low performance of rain-fed crop cultivation is not necessarily due to low physical potential, but largely to management

related issues. Thus, most smallholder crop farmers in sub Saharan Africa (SSA) engage in subsistence agriculture using traditional method of production probably because most modern technologies and innovations are not accessible to them.

Groundnut (*Arachis hypogaea* L.) is one of the most agronomical important food legumes grown in the northern parts of Ghana. Many producers and consumers depend on this leguminous crop for their livelihood and nutritional value. Groundnut is considered as the 3rd most important source of vegetable protein, 4th most important source of edible oil and 13th most important food crop in the world [6]. According to Girei et al. [7], groundnut seeds contain 50% high quality edible oil, 25% digestible protein and 20% carbohydrates. Its flour is use as ingredient in soup, confectioneries and pudding. Groundnut also provides high quality fodders and feeds for livestock, help in weeds control and soil water conservation. It also improves soil fertility by adding some organic matter into the soil and fixes atmospheric nitrogen into the soil.

In spite of the numerous uses of groundnut, availability of abundant land and human resources in the northern sector of Ghana; average yield per hectare for groundnut production has been on the decline over the years. It has been documented that the achievable yield of groundnut is 2.5 mt/ha with the average yield produced currently being 1.5 mt/ha [4]. The implication is that there is a gap for additional increase in output from the existing production level of groundnut. Therefore, it is of immense importance to overturn the preceding situation with the view to improving productivity and resource use among producers through the investigation of determinants of output and how various factors of production are allocated. It is against this background that the study intends to examine the resource-use-efficiency among groundnut producers in the Northern region of Ghana.

The remaining sections of this paper are presented as follows: Following this section is the presentation of the theoretical and empirical literature. The study area and the methodology are also described. The methodologies include a brief overview of the theoretical framework, sampling technique and data collection as well as the empirical model. The ensuing two sections present the results of the empirical findings and the conclusions as well as the policy recommendations.

Literature Review

The conventional notion of efficiency can be credited to the prominent works of Farrel [8] who suggested that; technical, allocative and economic efficiencies constitute the main components of efficiency. Farmers ability to achieve a maximum output given similar input levels measure their technical efficiencies whilst the optimum use of these inputs up to the level at which their marginal value of productivity is equal to the marginal factor cost is referred to as allocative efficiency. However, technical and allocative efficiencies are the main components of economic efficiency. There are two main ways of estimating efficiency of a firm, parametric and non-parametric [9-12]. The parametric can be categorized into two main components namely; the deterministic frontier models and the stochastic frontier models. The most common non-parametric approach is the data envelope analysis (DEA).

The parametric method of estimation involves the econometric modeling of production frontier. The parametric accounts for measurement errors in both output and stochastic elements by decomposing the effects of noise from the inefficiency effects. It also allows the conventional statistical test to be carried out. Unlike the parametric, the non-parametric method has the ability to measure efficiency of multi-output cases and requires no functional form to specify the relationship between inputs and outputs [10,13]. The main goals of resource use efficiency measurement is to find ways of increasing output per unit input and attaining desirable transfer of factors of production in other to raise our economic standard of living. Many authors have analyzed the efficiency of resource use in the agricultural sector by using farm level data from many parts of the world.

Taphee and Jongur [14] used Cobb-Douglas stochastic frontier analysis to analyze the productivity and technical efficiency of groundnut production in northern Taraba state, Nigeria. The empirical results showed that farm size, quantity of fertilizer, quantity of seeds and family labour were the key determinants of groundnut production in the study area. Moreover, the inefficiency component of the groundnut production in the study area was attributed to the age of the farmer, farmers contact with agricultural extension officers as well as the size of the family. Korir et al. [15] examined the determinants of Bambara groundnut production in Western Kenya using the stochastic frontier analysis found out that farmers farm size, amount of labour used and quantity of seeds were the major factors influencing groundnut production in the study area. The empirical results also indicated that, on the average, groundnut farms in the study area could increase their output by 61.58 percent using the same input level. That is, the study found the mean technical efficiency to be 38.42 percent. Shehu et al. [16] in analyzing the determinants of production and technical efficiency among yam farmers in Benue State, Nigeria showed that land area, fertilizer utilization, quantity of seeds and family labour were the major inputs influencing the production of yam in the study area. The empirical results also predicted yam farmer's efficiency to range between 67 percent and 99 percent with the mean efficiency level of 95 percent. Possible explanations to the variation of efficiency were attributed to factors such as educational attainment, membership of farmers association and household size. In a similar study conducted by Addai and Owusu [17] using stochastic transcendental (translog) production frontier to analyze the technical efficiency of smallholder maize farmers across the various agro-ecological zones of Ghana revealed that farm size and labour are the major factors influencing the production of maize in Ghana. Abdulai et al. [18] estimated the technical efficiency of maize production in Northern Ghana. Another study on the determinants of production and technical efficiency among cotton farmers in the Northern part of Ghana was conducted by Adzawala et al. [19]. The transcendental (translog) production frontier was used to estimate the production function. The empirical results revealed that farm size, labour and fertilizer utilizations are the main determinants of cotton production in the Northern part of Ghana.

Apart from the physical input factors that contribute to farmer's level of production and efficiency, there are socio-economic, demographics, institutional and environmental factors that affect farmer's efficiency level [20]. These factors include age, sex, marital status, household size, educational attainment, access to credit, engagement in off-farm income, land tenure system, membership of farmer based organizations, farmers contact with extension officers, etc.

Educational attainment enhances farmer's managerial and technical skills. It is hypothesized to increase farmer's ability to synthesized information and utilize the existing technologies to attain high efficiency levels [8]. However, Owour and Shem [21] have shown negative relationship between education and farmers efficiency levels. This was in variance with a study by Donkoh et al. [22] on efficiency of irrigated rice farmers in Northern Ghana. They observed that farmers with more years of formal education tend to be less technically inefficient than their counterparts with less years of formal education. It was also indicated in the same study that male farmers are more technically efficient than female farmers. A study by Diiro [23] to examine the impact of off-farm income on technical efficiency of maize farmers in Uganda concluded that farmers engaging in off-farm economic activities are less technically efficient than farmers with no off-farm income. Credit availability enables farmers to purchase adequate inputs and on timely manner and hence expected to increase farmers level of efficiency. The positive relationship between credit accessibility and farmers efficiency level was empirically supported by Chukwuji et al. [24] in analyzing the technical efficiency of farmers in Delta State, Nigeria. Asante et al. [25] in analyzing the effects of NERICA rice adoption on technical efficiency of rice farmers found household size, farmers contact with extension officers and access to road network to have positive influence on the farmer's level of efficiency.

Resource-use or allocative efficiency has been documented in many agricultural related literatures. Taru et al. [26] analyzed the economic efficiency of resource use among groundnut farmers in Nigeria. The empirical results indicated that the ratio of marginal value productivity to marginal factor costs for quantity of seeds and labour were greater than unity indicating under utilizations of these resources. Another study conducted by Maikasuwa and Ala [27] in determining profitability and resource-use

efficiency of yam production by women in Bosso local government area of Niger state, Nigeria, observed that fertilizer, labour and land were under-utilized. The under-utilization of fertilizer, land use and labour were also in line with a study conducted to examine the resource use efficiency of rice farmers in Jere local government area of Borno state, Nigeria by [28]. Another study conducted by Baiyegunhi et al. [29] in examining efficient allocation of resources among individual farms of different sizes in sorghum production in Kaduna state of Nigeria concluded that farm resources were inefficiently utilized for both small and large scales farmers. However, the study observed that quantity of seeds and fertilizers were under-utilized while the amount of labour employed was over-utilized.

2. METHODOLOGY

2.1 Study Area

The study was conducted in Tolon District in the Northern Region of Ghana, the largest region in the country in terms of land mass, constituting about 30% (70,390 km^2) of the total land area of the country. Tolon district shares borders' with North-Gonja to the west, Kumbugu district to the north, Central-Gonja to the south, whilst Tamale Metropolitan and Savelugu/Nanton District share the eastern boundaries with it. The district covers a total land mass of 2,741km^2 forming about 3.9% of the entire area of the Northern Region. The major economic activities are agriculture and its related works. The vegetation cover is basically Guinea Savanna interspersed with short drought resistant trees and grassland. The land is generally undulating with a number of scattered depressions. The soils are generally of the sandy loam type except in the low lands where alluvial deposits are found and support the cultivation of crops like groundnut, yam, cowpea, millet, sorghum, rice etc. The major tree species in the district include the sheanut, dawadawa, mango, which are economic trees and form an integral part of livelihood of its people.

2.2 Sampling Procedure and Data Collection

The information for the analysis was obtained from a cross-sectional primary data through an objective oriented structured questionnaire. The selection of the groundnut farm households followed a two-stage systematic random sampling. In the first stage, four groundnut producing communities (Tolon, Yogu, Nyankpala and Tali) were randomly selected from the list of major groundnut producing communities in the district. The second stage involved a random selection of one hundred and twenty (120) farm households across the four communities.

2.3 Analytical Technique

The study adopts the stochastic frontier analysis (SFA) and marginal value productivity-marginal factor cost (MVP-MFC) approach to achieve its objectives. The SFA was used to measure the ability of groundnut farmers to use a minimum quantity of inputs under a given technology to achieve a maximum level of output (Technical efficiency) while the MVP-MFC was used to measure their ability in achieving the best combination of different inputs in producing a given level of output considering the relative prices of these inputs (allocative efficiency). Thus, efficiency analysis in this study has been decomposed into two; technical and allocative efficiencies as documented in the literature of agricultural production efficiencies. Technical inefficiency arises when observed output from a given input mix is less than the frontier output. Allocative inefficiency arises when farmers fail to equalize their marginal returns with the true input market prices.

2.3.1 The SFA analysis

In stochastic frontier analysis, the farm is constrained to produce at or below the deterministic production frontier. The approach is preferred for efficiency studies in agricultural production due to the inherent stochastic nature of the agricultural systems. The stochastic frontier production function was independently proposed by Aigner et al. [30] and Meeusen and Van den Broeck [31]. The stochastic production function is defined by;

$$Y_i = f(x_i, \beta) + \varepsilon_i \text{ where } i = 1, 2, 3..., n \quad (1)$$

$$\varepsilon_i = v_i - u_i$$

Where Y_i represents the output level of the ith sample farm; $f(x_i, \beta)$ is a suitable function such as Cobb-Douglas or transcendental (translog) production functions, x_i is a vector of inputs for the ith farm and β is a vector of

unknown parameter. ε_i is an error term made up of two components: v_i is assumed to account for random effects on production associated with factors such as measurement errors in production and other factors which the farmer does not have control over and u_i is a non-negative error term associated with farm-specific factors, which leads to the ith farm not attaining maximum efficiency of production. Thus, u_i measures the technical inefficiency effects that falls within the control of the decision making unit.

Stochastic Frontier Analysis (SFA) has been used recently by many authors such as Donkoh et al. [22], Onumah et al. [32], Abdulai et al. [18], Danso-Abbeam et al. [33] and Onumah and Acquah [34]. The approach specifies technical efficiency of an individual farm as the ratio of the observed output to the corresponding frontier output conditioned on the level of inputs used by the farm. Technical inefficiency is therefore defined as the amount by which the level of production for the farm is less than the frontier output. Technical efficiency (TE) can be specified as;

$$TE = \frac{Y_i}{Y_i^*} = \frac{f(x_i; \beta) \cdot \exp(v_i - u_i)}{f(x_i; \beta) \cdot \exp(v_i)} = \exp(-u_i) \quad (2)$$

Where Y_i or $f(x_i, \beta) \cdot \exp(v_i - u_i)$ is the observed output and Y_i^* or $f(x_i, \beta).\exp(v_i)$ is the unobserved output. According to Battese and Coeli [35], the error term v_i is assumed to be identically, independently and normally distributed with zero mean and a constant variance, $N(0, \sigma_V^2)$. The error term u_i is also assumed to be distributed as truncation of the normal distribution with mean u_i and variance $N(u_i; \sigma_u^2)$ such that the inefficiency error term can be explained by exogenous variables specified as;

$$u_i = Z\delta_i + W_i \quad (3)$$

Where Z_i is a $(1 \times M)$ vector of explanatory variables, δ_i is an $(M \times 1)$ vector of unknown parameters to be estimated; and W_i are unobservable random variables. In this study, a single -stage maximum likelihood approach was used to estimate the technical efficiency levels of the groundnut farmers and the determinants of technical inefficiency simultaneously. This simultaneous estimation approach ensures that the assumption of identical distribution of the error term u_i is not violated. The maximum likelihood estimates of the stochastic frontier model provide the estimates of β and the gamma (γ), where the gamma explains the variation of the total output from the frontier output. The gamma estimate is specified as, $\gamma = \dfrac{\sigma_u^2}{\sigma^2}$, where γ lies between zero and one $(0 \leq \gamma \leq 1)$, σ_u^2 is the variance of the error term associated with the inefficiency and σ^2 is the overall variation in the model specified as the sum of the variance associated with the inefficiency (σ_u^2) and that associated with random noise factors (σ_v^2). Thus; $\sigma^2 = \sigma_u^2 + \sigma_v^2$.

The closer the value of the gamma (γ) is to one (1), the more the deviation of the observed output from the deterministic output which is as a result of inefficiency factors. However, if the value is closer to zero, then deviations are as a result of random factors and if the value lies between one (1) and zero (0), then deviations are as a result of both inefficiency and random factors.

2.3.2 Allocative efficiency analysis

In other to evaluate the extent to which groundnut farmers in the study area are putting their resources into efficient use, the study adopts the marginal-value-productivity-marginal-factor- cost analysis. This method has been used by many authors (Sienso et al. [36], Gani and Omonona [37], Oladeebo and Ambe-Lamidi [38]), where the marginal value productivities (MVPs) for each input used were computed and such computed MVPs were then compared with their respective acquisition cost, marginal factor cost (MFC).

For transcendental logarithmic (translog) production function, we estimate the factor elasticities (β_i) and the marginal physical products from the OLS estimates of the translog production function with respect to each input

used. We then use the β_i and the MPP to compute MVP and RUE as shown in equation [4, 5, and 6] below.

$$MVP = MPP \cdot P_Y \qquad (4)$$

Where

$$MPP = \beta_i \cdot {Y_i}\big/{X_i} \Rightarrow MVP = \left(\beta_i \cdot {Y_i}\big/{X_i} \right) \cdot P_Y \quad (5)$$

Where Y_i is the mean groundnut output of the ith farmer, X_i is the mean input used and P_Y is the price of output. The resource-use-efficiency (RUE) of each of the measurable input used in groundnut production was computed by the ratio of the marginal value of productivity (MVP) to that of the marginal factor cost (MFC). Thus,

$$RUE = \frac{MVP}{MFC} \qquad (6)$$

Where RUE denotes resource-use-efficiency and MFC represent the price of the measurable factor inputs at their geometric means.

Decision rule

i. $RUE = 1$, implies that resources are used efficiently by groundnut farmers in the study area.
ii. $RUE > 1$, implies resources are under-utilized and increasing the rate of use of that resource will help increase productivity.
iii. $RUE < 1$, implies resources are over-utilized and reducing the rate of use of that resource will help improve productivity.

2.3.3 The empirical model specification

In estimating the stochastic production frontier function of groundnut production in the study area, we used the transcendental (translog) production function developed by Christensen et al. [39], after a preliminary test of hypotheses had suggested that Cobb-Douglas production function was inadequate representation of the data. The transcendental logarithmic (translog) production function had been used consistently in many recent agricultural efficiency related studies such as [22,32,36].

Following Battese and Coeli [35], the transcendental stochastic frontier model can be expressed as;

$$\ln Y = \beta_0 + \sum_{j=1}^{4} \beta_j \ln X_{ji} + \sum_{j\le}^{4} \sum_{k=1}^{4} \beta_{jk} \ln X_{ji} \ln X_{ki} + V_i - U_i \quad (7)$$

Where Y, X_1, X_2, X_3 and X_4 denotes the output level (kilograms), farm size (acres), quantity of seeds sowed (kilograms), quantity of herbicides (liters) and labour use (man-days) respectively.

The empirical model for the inefficiency model can also be specified as;

$$U_i = \delta_0 + \delta_1 Z_1 + \delta_2 Z_2 + \delta_3 Z_3 + \delta_4 Z_4 + \delta_5 Z_5 + \delta_6 Z_6 + \delta_7 Z_7 + \delta_8 Z_8 + W \quad (8)$$

Where Z_1, Z_2, Z_3, Z_4, Z_5, Z_6, Z_7 and Z_8 are sex of the farmers (categorized as 1 for male and 0 for female), age of the farmer, number of years in formal education, marital status, number of years in farming, household size, farmer belonging to any farmer-base-organization and the number of extension visits respectively. W_i is the error term and δ_i is the vector of parameters to be estimated.

2.3.4 Specification of hypotheses

In estimating the stochastic frontier model for groundnut farmers in the study area, three main null hypotheses were conducted to examine the appropriateness of the specified model used, the absence of inefficiency and the significance of socio-economic factors in explaining inefficiency among groundnut farmers. The three hypotheses are presented as follows;

1. $H_0 : \beta_{ji} = 0$ The coefficients of the square values and the interaction terms in the translog model sum up to zero
   $H_1 : \beta \ne 0$ The coefficients of the square values and the interaction terms in the translog model do not sum up to zero

2. $H_0 : \gamma = \delta_0 = \delta_1 .... \delta_8 = 0$
   There are no inefficiency effects
   $H_1 : \gamma = \delta_0 = \delta_1 ..... \delta_8 \ne 0$
   There are inefficiency effects

3. $H_0 : \gamma = 0$
   Inefficiency effects are stochastic

$H_1 : \gamma \ne 0$
Inefficiency effects are non-stochastic

These hypotheses were tested by the use of the generalized likelihood-ratio test statistic specified as;

$$LR(\lambda) = -2[\{\ln L(H_0)\} - \{\ln L(H_1)\}] \quad (9)$$

Where $L(H_0)$ and $L(H_1)$ are the likelihood functions under null and alternate hypotheses respectively. If the given null hypothesis is true, then the test statistic (λ) has a chi-square distribution of degree of freedom which is equal to the difference between the estimated parameters under (H_1) and (H_0). However, if the null hypothesis involves $\gamma = 0$, then the asymptotic distribution involves a mixed chi-square distribution [40].

3. RESULTS AND DISCUSSION

3.1 Demographic Characteristics of the Sampled Farm Households

Tables 1 and 2 present the descriptive statistics of the sampled farm households. Out of the 120 household heads interviewed, 90 were females and 30 were males representing 75% and 25% respectively. On the average, there were 12 people per household and the average age of a household head was 35.5 years as indicated in Table 2 below. This indicates a relatively economic active adult population who has the ability to work very hard to increase the productivity level of groundnut in the region and the country as a whole if well aggravated. This is in contrast to a study by (FASDEP II) [41] that stated that Ghanaian farming population is generally old, as farming does not seem to attract the younger population. Moreover, majority (88.3%) of the farming population had primary education (6 years) with only 11.7% being able to attain junior high school and senior high school. Table 2 further indicated that, on the average farmers had been in groundnut farming for about 29 years. In Ghana, over 92% of the farming population farm on small scale [22].

This situation is not different from the one pertaining in the study area, considering the fact that farm size ranges between 1-9 acres with the mean farm size of 2.8 acres. From Table 2, the average total output of groundnut is about 82 kg while mean quantity of groundnut seeds sowed per plot (seed density) was about 44 kg.

3.2 Test of Hypotheses

As indicated in section 2.3.4, the study seeks to test three main hypotheses. The tests of these hypotheses are presented in Table 3. Table 3 indicates that the decision to use Cobb-Douglas frontier function was rejected in favour of transcendental logarithmic (translog) frontier function since the generalized likelihood-test statistic is significantly different from zero. This means that the results from the translog model are more accurate and adequate representation of the data, given the assumptions of the frontier model. The result of the second hypothesis revealed that the frontier production function was more appropriate to fit the data than the average response production function. Moreover, findings from the third hypothesis suggest that inefficiency effects are present in the model so the decision to exclude them was discarded.

3.3 Groundnut Stochastic Frontier Production Function Analysis

Results for the estimation of stochastic frontier production of groundnut in the study area are presented in Table 4 below. The estimated sigma squared of 0.24 indicates a "good fit" and the appropriateness of the specified distributional assumption of the composite error term rather than the average response specification. Gamma measures the level of inefficiency in the variance parameter, that is, the difference between the frontier output and the observed output. The estimated gamma value of 0.83 indicates that 83% variation in groundnut output was due to inefficiency in input use and other farm practices whilst 17% of the deviations of the actual output from the frontier output came from random factors. Some of these factors could be in the form of pest and disease infestation, unfavorable weather conditions and statistical errors in data collection and measurement.

Table 4 also measures groundnut productivity in terms of output elasticities. Output elasticities responded positively to the amount of labour and quantity of seeds used and negatively to the area of farm land allocated to groundnut farming. The results further demonstrated that a percentage increase in the amount of labour and quantity of seeds sown increase groundnut output by 7.84% and 5.84% respectively. These findings are consistent with the results of similar studies

conducted recently by Taphee and Jongur [14] and Ani et al. [42]. In the case of farm size, a percentage increase in farm land allocated to groundnut farming decreases output by 7.80%. The inverse relationship between output and farm size could partly be attributed to poor farm management practices. For example, farmers may not effectively combine land with other factors of production such as labour and seeds as they expand their farm lands. This is in contrast to some other studies in other parts of Ghana and other African countries which indicated that, farm size is an increasing function of output [43,36,23,44]. However, some other studies in Ghana by Donkoh et al. [22] and Adzawala et al. [19] found similar results in relation to rice and cotton respectively.

The squared variables in the translog stochastic frontier function indicate the effect of continuous use of that variable on output. The interaction terms indicate a complementarity or substitutability of the inputs employed on the farm. A significant positive coefficient of interaction term means the two variables are complements whilst a significant negative term means the two variables are substitutes. The results indicated that continuous use of quantity of seeds have a negative significant influence on groundnut output whilst the continuous use of land, labour and herbicides have no significant effects on output. Moreover, Table 4 indicates that quantity of labour employed has a significant complementarity with quantity of seeds sown. Similarly, farm land allocated to groundnut farming is complementary to quantity of herbicides used on the farm.

3.4 Level of Technical Efficiencies

Table 5 shows a considerable variation of efficiency index across groundnut farms in the study area. The predicted farm efficiency levels ranged between 20.04% and 98.65%. The mean technical efficiency was estimated to be about 84%. This implies that, the average groundnut farmer in the study area produces about 84% of the potential output given the current technology available. That is, groundnut farmers in the study area produce at a level below 16% of the frontier output. Thus, in the short run, there is enough room for groundnut farmers to increase their production by 16% by adopting new technologies, farm practices and efficient combination of inputs. Similar results have been documented by Taphee and Jongur [14] who estimated the mean technical efficiency of groundnut farmers in Northern Taraba State of Nigeria to be 97%. However, this is far higher compared with the results obtained by Ani et al. [42] who estimated the mean technical efficiency of groundnut farmers in Nigeria to be 3.78%. The findings further revealed that majority (73.33%) of the farmers operated with technical efficiency level of 70% and above, whilst only few (14.16%) had technical efficiency level of 50% and below.

3.5 Determinants of Technical Inefficiency

The results presented in Table 6 identify the sources of variation in technical efficiency estimates. The variables used in the technical inefficiency model are the determinants of technical inefficiency rather than efficiency. This implies that a positively signed variable reduces technical efficiency level whilst a negatively signed variable increases technical efficiency level.

The coefficient of education was negatively signed and significant at 10% indicating that farmers who had more years of formal education were more technically efficient than their counterparts with less years of formal education. Recent studies by Mapemba et al. [45] had established that education is a variable that sharpens farmers' managerial skills and hence, improves their efficiency level. Similar result was documented by Asante et al. [46]. Farmer's level of experience also had a negative sign signifying an increasing function of technical efficiency. This could partly be attributed to the fact that farmers with longer years in farming are able to draw from their past experiences to suit their farming practices and conditions. The negative and significant coefficient of frequency of extension visit suggests that farmers who have the opportunity to frequent extension services reduce their level of technical inefficiency. Thus, increasing the frequency of extension contacts with farmers by a unit reduces their level of inefficiency by 0.27 units. Other researchers such as Parkh et al. [47] and Seyoum et al. [48] have documented similar findings.

The coefficient of household size exhibits a negative function of farmer's efficiency level as it is positively signed and significant at 5%. The size of the coefficient indicates that farmers with larger family size are less technically efficient than their counterparts with smaller family size. Thus, increasing farm household size by a percentage increases farmer's inefficiency level

by 0.155%. This is contrary to our a priori expectation that farmers with larger family size will have more family labour which may in turn increase their level of efficiency. The possible explanation is that farming and for that matter groundnut cultivation is not the only economic activity performed by the farmers in the study area; therefore household labour may allocate more of their time to other farming and non-farming activities. Ani et al. [42] also found household size to have positive influence on technical inefficiency of groundnut farmers in Benue State, Nigeria.

The coefficient of farmer-base-organization has negative significant effect on farmer's technical efficiency level. That is, members of farmer-base-organization are less technically inefficient than non-members of farmer-base-organization. This could be due to the fact that farmer-base-organization members receive input and support services from many donors and NGO's. The results of the study oppose to the one conducted by Addai et al. [49] who documented that there is no significant difference in terms of technical efficiency between members and non-members of farmer-base-organization.

In contrast to other studies like Addai et al. [49] and Donkoh et al. [22], age and sex were estimated in this study to have no significant influence on inefficiency level of groundnut production. While Addai et al. [49] concluded that older farmers are less technically inefficient; Donkoh et al. [22] stressed that male farmers are technically more efficient than their female counterparts.

3.6 Resource-use-efficiency Estimation

In this study, resource-use-efficiency of groundnut farmers was measured by the ratio of the marginal-value-productivity (MVP) of each input used to their respective prices. Thus, the marginal-value-productivity is the yardstick for judging how resources are allocated. Inputs are said to be well allocated under pure competitive condition when there is no divergence between their MVP and their unit price. The MVP for each input was determined by multiplying the MPP of each input by the mean price of the groundnut output. In this study, we measured resource-use-efficiency for three inputs namely; seeds, labour and herbicides. Resource-use-efficiency for land was not considered because land is a fixed input and its adjustment depends on long term profitability. The results of the resource-use-efficiency estimations are presented in Table 7 below. The results suggested that all the variable inputs under consideration were not used efficiently.

The ratio of *MVP/MFC* for labour and quantity of seeds were less than unity indicating over-utilization of these inputs, hence increasing the quantity of labour and seeds usage would decrease output and profit level. However, quantities of herbicides were under-utilized with an efficiency score of 3.183. The under-utilization of herbicides may be due to the high cost of herbicides in the study area.

Table 1. Distribution of respondents by demographic characteristics

Variable	Range	Respondents	Percentages
Age	19-29	38	31.7
	30-39	44	36.6
	40-49	22	18.3
	50-59	10	8.3
	60+	6	5.0
	Total	120	100
House hold size	1-9	56	46.7
	10-19	37	30.8
	20-29	23	19.2
	30-39	3	2.5
	40-49	1	0.8
	Total	120	100
Education	1-6	106	88.3
	7-12	14	11.7
	Total	120	100
Male		30	25
Female		90	75

Source: Field Survey, 2014

Table 2. Descriptive statistics of variables

Variable	Minimum	Maximum	Mean	Std. deviation
Sex	0	1	0.73171	0.44488
Marital status	0	1	0.94309	0.23262
Age (years)	19	75	35.46341	10.89239
Educational level (years)	0	12	3.00293	2.76423
Experience (years)	17	43	28.82927	6.8577
Household size	1	40	11.7561	7.67938
Farmer base organization	0	1	0.52032	0.50163
Extension services	0	1	0.07317	0.26148
Output (bowls; 2.5kg/bowl)	25.7	520	82.83902	106.1756
Farm size (acres)	1	9	2.76423	1.50317
seeds (kg)	13	214.5	43.84553	24.63914
Herbicides (litres)	1	9	2.743902	1.499133
Labour (man-days)	28	400	132.2358	88.22754

Source: Field survey, 2014

Table 3. Results of hypotheses tested

Test type	Test statistic	P-value	Decision rule
Functional form test	16.80	0.079	Reject H_0: Translog is appropriate
Frontier test	25.06	0.016	Reject H_0: Frontier is appropriate
Inefficiency test	28.42	0.0001	Reject H_0: Inefficiency effects are present in the model

Source: Field Survey, 2014

Table 4. Maximum likelihood estimates of the stochastic frontier production function

Variable	Parameter	Coefficient	Std error	P-value
Constant	β_0	-8.039284	4.544821	0.077
Labour	β_1	7.844772	4.700776	0.095
Herbicides	β_2	1.120254	1.529197	0.464
Seeds	β_3	5.835798	2.658528	0.028
Farm size	β_4	-7.803223	4.331920	0.072
(labour)(labour)	β_{11}	-0.459161	0.280501	0.102
(Herb)(Herb)	β_{22}	-1.631556	1.294464	0.208
(Seeds)(Seeds)	β_{33}	-1.660833	0.505938	0.001
(Farm size)(Farm size)	β_{44}	-1.475826	1.796383	0.411
(Labour)(Herbicides)	β_{12}	-0.304701	1.103999	0.783
(Labour)(Seeds)	β_{13}	0.925123	0.420090	0.028
(Labour)(Farm size)	β_{14}	0.609402	1.130474	0.590
(Herbicides)(Seeds)	β_{23}	-2.681500	1.657997	0.106
(Herbicides)(Farm size)	β_{24}	3.098044	1.714294	0.071
(Seeds)(Farm size)	β_{34}	2.420524	1.584224	0.127
Sigma- squared		0.2447		
Gamma		0.8305		
Log likelihood function		-24.039		

Source: Field Survey, 2014. *** = 1% significance level, * = 10% significance level

Table 5. Frequency distribution of technical efficiency index

Efficiency score	Percentage (%)	Frequency	Cumulative
20 - 30	3.33	4	4
31 - 40	5.00	6	10
41 - 50	5.83	7	17
51 - 60	12.51	15	32
61 - 70	15.00	18	50
71 - 80	13.33	16	66
81 - 90	20.00	24	90
91 - 100	25.00	30	120
Maximum	100	120	
Mean	98.65		
Minimum	83.89		
	20.04		

Source: Field Survey, 2014

Table 6. Determinants of technical inefficiency in groundnut production

Variable	Parameter	Coefficient	Std Error	P-value
Constant	δ_0	0.887395	0.3966230	0.027
Sex	δ_1	0.003213	0.0042131	0.447
Age	δ_2	0.011479	0.0090036	0.205
Education	δ_3	-0.004783	0.0025706	0.065*
Marital status	δ_4	0.001923	0.0039016	0.623
Experience	δ_5	-0.653460	0.9199170	0.000***
Household size	δ_6	0.154586	0.0679238	0.025**
Membership of FBO	δ_7	-0.113706	0.0579535	0.052**
Contact with extension service	δ_8	-0.272325	0.1397614	0.054**

Source: Field survey, 2014. *** = 1% significance level, * = 10% significance level

Table 7. Resource-use-efficiency estimates

Variable	MFC(GH¢)	MVP	RUE = MVP/MFC	Decision rule
Labour	4.5/man-day	0.6205	0.865	Over-utilization
Seeds	10/22.5kg	0.308	0.308	Over-utilization
Herbicides	10/22.5kg	3.183	3.183	Under-utilization

Source: Field survey, 2014

4. CONCLUSIONS AND RECOMMENDATIONS

The focus of this study was to estimate the resource-use-efficiency in the Northern region of Ghana. The results indicated that amount of labour and quantities of seeds have positive effects on output of groundnut in the study area. However, farm size allocated to groundnut farming was estimated to have a negative effect on output. The mean technical efficiency was estimated to be 83.89% with majority of groundnut farmers achieving efficiency score of 73% and above. The sources of variation in the technical inefficiencies of farmers include education, experience, household size and frequency of extension visits. The study further demonstrated that groundnut farmers in the study area exhibited positive decreasing return to scale implying that, increase in the factors of production produce less than proportionate increase in output. The resource-use-efficiency analysis had indicated that none of the variable inputs employed by the farmers was efficiently utilized. The study therefore recommends a farm level policy directed towards the stimulation of extension work through motivation to give the rural farm households the needed training on farm management to improve productivity.

COMPETING INTERESTS

Authors have declared that no competing interests exist.

REFERENCES

1. Government of Ghana. Budget Statement and Economic Policies of the Budget for a brighter future; 2009. Available:http//www.mofep.gov.gh
2. Breisinger C, Diao X, Thurlow J, Al-Hassan RM. Potential impacts of green revolution in Africa – The case of Ghana, Paper presented at the 27th IAAE Conference, Beijing; 2009.
3. Asekenye C. An Analysis of Productivity Gaps among Smallholder Groundnut Farmers in Uganda and Kenya. Master's Thesis: Paper 323; 2012. Available:http://digitalcommons.uconn.edu/gs_theses/323
4. Facts and Figures, Statistics, Research and Information Directorate, Ministry of Food and Agriculture (MoFA). Accra, Ghana; 2012.
5. Rockström J, Barron J, Fox P. Water productivity in rain-fed agriculture: Challenges and opportunities for

smallholder farmers in drought-prone tropical agro-ecosystems; in Kijne JW, Molden D, Barker R. Water productivity in agriculture: Limits and opportunities for improvement. Comprehensive Assessment of Water Management in Agriculture Wallingford, UK, CABI International. 2003;(1):145-162.

6. Taru VB, Kyagya IZ, Mshelia SI. Profitability of groundnut production in michika local government area of adamawa state, Nigeria. Journal of Agricultural Science. 2010;1(1):25-29.

7. Girei AA, Duana Y, Dire B. An Economic Analysis of Groundnut (*Arachis hypogea*) Production in Hong Local Government Area of Adamawa State, Nigeria. Journal of Agricultural and Crop Research. 2013; 1(6): 84-89.

8. Farrell MJ. The measurement of productive efficiency. Journal of Royal Statistical Society. 1957;3:253-290.

9. Battese GE. Frontier production functions and technical efficiency: A survey of Empirical Applications in Agricultural Economics. Agricultural Economics. 1992;7:185-208.

10. Coelli T, Rao DSP, O"Donnell CJ, Battese GE. An Introduction to Efficiency and Productivity Analysis. Springer. New York, USA; 2005.

11. Ray SC. Data Envelopment Analysis: Theory and Techniques for Economics and Operation Research. Cambridge University Press, New York; 2004.

12. Zhu J. Quantitative models for performance evaluation and benchmarking: Data Envelopment Analysis with Spreadsheets and Excel Solver. Kluwer Academic Publishers, Boston; 2003.

13. Battese GE, Broca SS. Functional forms of stochastic frontier production functions and models for technical inefficiency effects: A comparative study for wheat farmers in Pakistan. Journal of Productivity Analysis. 1997;8:395-414.

14. Tapheel GB, Jongur AAU (2014) Productivity and Efficiency of Groundnut Farming in Northern Taraba state, Nigeria. Journal of Agriculture and Sustainability. 2014;5(1):45 – 56.

15. Korir MK, Serem AK, Sulo TK, Kipsat, MJ. A Stochastic Frontier Analysis of Bambara Groundnut Production in Western Kenya. 18th International Farm Management Congress, Methven, Canterbury, New Zealand. 2013;3:74 – 80.

16. Shehu JF, Tashikalma AK, Gabdo GH. Efficiency of resource use in small scale rain-fed upland rice production in Northwest Agricultural Zone of Adamawa State, Nigeria. 9th Annual National Conference of Nigeria. Association of Agricultural Economics (NAAE) Held at ATBU, Bauchi, Nigeria; 2007.

17. Addai KN, Owusu V. Technical Efficiency of Maize Farmers across the Various Agro Ecological Zones. Journal of Agriculture and Environmental Sciences. 2014;3(1): 149 – 172.

18. Abdulai S, Nkegbe PK, Donkoh SA. Technical efficiency of maize production in Northern Ghana. African Journal of Agricultural Research. 2013;8(43);5251-5259.

19. Adzawla W, Fuseini J, Donkoh SA. Estimating technical efficiency of cotton production in Yendi Municipality, Northern Ghana. Journal of Agriculture and Sustainability. 2013;4(1):115-140.

20. Kumbhakar SC, Bhattacharya A. Price distortion and resource use efficiency in indian agriculture: A Restricted Profit Function Approach. Review of Economics and Statistics. 1992;74:231-239.

21. Owuor G, Shem OA. What Are the Key Constraints in Technical Efficiency of Smallholder Farmers in Africa? Empirical Evidence from Kenya. A Paper Presented at 111 EAAE- IAAE Seminar Small Farms: Decline or persistence' University of Kent; 2009.

22. Donkoh SA, Ayambila S, Abdulai S. Technical Efficiency of Rice Production at the Tono Irrigation Scheme in Northern Ghana. American Journal of Experimental Agriculture. 2013;3(1):25-42.

23. Diiro MG. Impact of Off-farm Income on Agricultural Technology Adoption Intensity and Productivity. Evidence from rural maize farmers in Uganda. International Food Policy Research Institute, Uganda Food Support Programme. Working Paper 11; 2013.

24. Chukwuji CO, Inoni OE, Ike PC. Determinants of technical efficiency in Gari in Delta State, Nigeria. Journal of Central European Agriculture. 2007;8(3):327-336.

25. Asante BO, Wiredu AN, Martey E, Sarpong DB, Mensah-Bonsu A. NERICA Adoption and Impacts on Technical Efficiency of Rice Producing Households in Ghana:

Implications for research and development. American Journal of Experimental Agriculture. 2014;4(3):244-262.

26. Taru VB, Kyagya IZ, Mshelia SI, Adebayo EF. Economic Efficiency of resource use in groundnut production in Adamawa State of Nigeria. World Journal of Agricultural Sciences. 2008;4(s):896 – 900.

27. Maikasuwa MA, Ala AL. Determination of Profitability and Resource-use Efficiency of Yam Production by Women in Bosso Local Government Area of Niger State, Nigeria. European Scientific Journal. 2013;9(16): 198 – 205.

28. Tijani BA, Abubakar M, Benisheik KM, Mustapha AB. Resource Use Efficiency in Rice Production in Jere Local Government Area of Borno State, Nigeria. Nigerian Journal of Basic and Applied Science. 2010;18(1):27 – 34.

29. Baiyegunhi LJS, Fraser GCG, Adewumi MO. Resource Use Efficiency in Sole Sorghum Production in Three Villages of Kaduna State, Nigeria. African Journal of Agricultural Research. 2010;5 (3):172-177.

30. Aigner DJ, Lovell CA, Schmidt P. Formulation and estimation of stochastic frontier production function Model. Journal of Econometrics. 1977; 1(1): 21-37.

31. Meeusen W, Van den Broeck J. Efficiency Estimation from Cobb–Douglas Production Functions with Composed Error. International Economic Review. 1977; 18(2):435–443.

32. Onumah JA, Al-Hassan RM, Onumah EE. Productivity and technical efficiency of cocoa production in Eastern Ghana. Journal of Economics and Sustainable Development. 2013;4(4):106 – 118.

33. Danso-Abbeam G, Aidoo R, Agyemang KO, Ohene-Yankyera K. Technical Efficiency in Ghana's Cocoa Industry: Evidence from Bibiani -Anhwiaso-Bekwai District. Journal of Development and Agricultural Economics. 2012;4(10):287-294.

34. Onumah EE. Acquah HD. frontier analysis of aquaculture farms in the Southern Sector of Ghana. World Applied Sciences Journal. 2010; 9(7): 826-835.

35. Battese GE, Coelli TJ. A Model for Technical Inefficiency Effects in a stochastic frontier production function for panel data. Empirical Economics. 1995;20(2):325-32.

36. Sienso G, Asuming-Brempong S, Amegashie DPK. Estimating Efficiency of

Maize Farmers in Ghana. Asian Journal of Agricultural Extension, Economics & Sociology. 2014; 3(6): 705-720.

37. Gani BS, Omonona BT. Resource Use efficiency among small - scale. irrigated maize producers in Northern Taraba State of Nigeria. Journal of Humun Ecology. 2009; 28(2):113-119.

38. Oladeebo JO, Ambe-Lamidi AI. Profitability, Input Elasticities and Economic Efficiency of Poultry Production among Youth Farmers in Osun State, Nigeria. International Journal of Poultry Science. 2007; 6(12):994-998.

39. Christensen LR, Jorgensen DW, Lau LJ. Transcendental Logarithmic Production Frontiers. Review of Economics and Statistics. 1973;55:28-45.

40. Coelli TJ. Estimators and Hypothesis Tests for a Stochastic Frontier Functions: A Monte Carlo Analysis. Journal of Productivity Analysis. 1995;6:247-268.

41. Food and Agriculture Sector Development Policy (FASDEP II). Ministry of Food and Agriculture, MoFA. Accra, Ghana; 2007.

42. Ani DP, Umeh JC, Weye EA. Profitability and Economic Efficiency of Groundnut Production in Benue State, Nigeria. African Journal of Food, Agriculture, Nutrition and Development. 2013;13(4): 8091 – 8105.

43. Suresh A, Reddy TRK. Resource-use Efficiency of Paddy Cultivation in Peechi Command Area of Thrissur District of Kerala: An Economic Analysis. Agricultural Economics Research Review. 2006;19:159-171.

44. Nkegbe PK. Technical Efficiency in Crop Production and Environmental Resource Management Practices in Northern Ghana. Environmental Economics. 2012;3(4):43 – 51.

45. Mapemba DL, Maganga MA, Mango N. Farm household production efficiency in Southern Malawi: An Efficiency Decomposition Approach. Journal of Economics and Sustainable Development. 2013;4(3):236 – 245.

46. Asante BO, Villano RA, Battese GE. The effect of the adoption of yam minisett technology on technical efficiency of yam farmers in the forest-savanna transition zone of Ghana. African Journal of Agricultural and Resource Economics. 2014;9(2):75-90.

47. Parikh A, Shah K. Measurement of technical efficiency in the North-west Province of Pakistan. Journal of

Agricultural Economics. 1995;4(5):133-137.

48. Seyoum ET, Battese GE, Fleming EM. Technical Efficiency and Productivity of Maize Producers in Eastern Ethiopia: A study of Farmers within and outside the Sasakawa Global 2000 project. Agricultural Economics. 1998;(19):341-348.

49. Addai KN, Owusu V, Danso-Abbeam G. Effects of farmer – Based- organization on the technical efficiency of maize farmers across Various Agro - Ecological Zones of Ghana. Journal of Economics and Development Studies. 2014;2(1):141–161.

2

Performance and Gut Morphometry of Broiler Fed Maize Based Diets Supplemented with Charcoal and Honey as Anti-aflatoxin

O. A. Adebiyi[1*], U. V. Okolie-Alfred[1], C. Godstime[1] and O. A. Adeniji[1]

[1]Department of Animal Science, Faculty of Agriculture and Forestry, University of Ibadan, Ibadan, Oyo State, Nigeria.

Authors' contributions

This work was carried out in collaboration between all authors. Authors OAA and UVO designed the study, authors UVO and CG carried out the feeding trial and handled literature searches. Author Adeniji wrote the first draft. Author OAA performed the statistical analyses. All authors read and approved the final manuscript.

Editor(s):
(1) Zhen-Yu Du, School of Life Science, East China Normal University, China.
(2) Daniele De Wrachien, Faculty of Irrigation and Drainage, State University of Milan, Italy.
Reviewers:
(1) Anonymous, Botswana.
(2) Anonymous, Egypt.
(3) Anonymous, Pakistan.

ABSTRACT

The study was conducted to assess The effect of charcoal and honey on the zoo technical performance and gut morphometry of broiler birds fed naturally aflatoxins contaminated maize based diets in a comparative study was carried out. In a completely randomised design, 240 one-week old Arbor Acre broilers were distributed randomly to six dietary treatments with four replicates of ten birds per replicate. The treatments were as follows: T1= Normal diets (diet formulated with normal maize) (positive control, with 15 ppb AF); T2= Rejected maize diets (negative control, with 32 ppb AF); T3=Positive control plus 2% charcoal; T4=Rejected maize diets plus 2% charcoal; T5=Positive control plus 2% honey; T6=Rejected maize diet plus 2% honey. Feeding and provision of water were supplied ad-libitum. On the 42th day, nine birds per treatments were slaughtered for gut morphometric attributes (villus height, crypt depth, villus width and villus height to crypt depth ratio) of duodenum, jejunum and ileum part of the gut. The zoo technical performance of broiler in the experiment were not significantly different (P<0.05) from all the treatments despite having different feed conversion ratio. Results from the gut morphology showed that the least villus height of birds from duodenum was recorded on AFL (180 x 10^2 mm) an indication of effect of aflatoxin. On the ileum, AFL (138 x 10^2 mm), CTL-Ho (118 x 10^2 mm) and AFL-Ho (116 x 10^2 mm) of villus height of birds were not significantly ($P<0.05$) different from each other. However, the histopathology of liver, kidney and bursa of fabricius showed healing power of honey as no visual lesions was seen on the slides of the organs prepared. In conclusion, 2% charcoal–to-aflatoxins-contaminated feed was more effective than 2% honey.

Keywords: Growth performance; gut morphology; organ histopathology; adsorbent.

1. INTRODUCTION

Recent trend in research for animal nutritionist has been directed to solving the issue of contamination of animal feeds with mycotoxins since they pose serious threat to both humans and animals (Hussein and Brasel [1]; Wu and Munkvold [2]; Zhang and Cauper [3]). Mycotoxins are secondary toxic metabolites of fungi growth (*Aspergillus flavus, A. parasiticus and A. nominus*) on grains, forages and even on dead woods under favourable environmental conditions Dutta and Das [4]. Mycotoxin contamination of poultry feeds have been the second major stumbling block in feed industry after increasing price of convectional feedstuffs Sundu et al. [5]. It has been reported that 25% world's grain was contaminated by mycotoxins as revealed by FAO survey Devegowda and Murthy [6,7]. Among over 300 of mycotoxins identified aflatoxins (B$_1$ B$_2$ G$_1$ and G$_2$) are the most studied and the main concern for animal nutritionist as well as its residue in animal products like milk, muscle i.e. AFM_1 and AFM_2 Yiannikouris and Jouany [8].

Several approaches have been employed in the research of aflatoxin decontamination. Out of the chemical, physical and biological means of detoxification of aflatoxin Lopez-Garcia and Park [9]; Sinha [10] studies carried out, adsorbent-based studies have been reported to be effective in removing aflatoxin from contaminated feed and minimise the toxicity of aflatoxin in poultry Ibrahim et al. [11]. Of all the commercially available adsorbing agent in the market, zeolite Mazzo et al. [12], bentonites Rosa et al. [13]; Pasha et al. [14] and Clinoptilolite (CLI) Oguz and Kurtoglu; Oguz et al. [15,16] were the most preferred as a result of their capacity to bind effectively with AF as well as the reduction in the effect of AF-absorption from the gastrointestinal tract. However, the use of clays have been reported to be toxisorbent (mycotoxin specific) and binds to certain essential nutrients needed for animal growths Chestnuts, [17]. The use of charcoal has been reported to be enterosorbent

i.e. has the capacity to binds to several mycotoxins Whitlow [18] due to its large surface area with negatively charged site. Edrington et al. [19] have shown that charcoal was used as a binder in broiler. Wellford et al. [20] reported that honey had an antifungal effect against *A. flavus* and *A. parasiticus* and an even stronger anti-aflatoxigenic effect. Honey contains some unknown substances that make it serve as a therapeutic agent against sores or lesions. This makes it a good ameliorating agent against the effects of AF on the organs of the animals

2. MATERIALS AND METHODS

2.1 Aflatoxin Quantification Analysis

High performance liquid chromatography (HPLC) was used to identify and quantify the presence of aflatoxins {B$_1$ B$_2$ G$_1$ and G$_2$} and other mycotoxins if presents. This analysis was carried out in the pathology unit of the International Institute of Tropical Agriculture (IITA), Ibadan. After extraction, the sample was made up to 1 ml and 4ul spotted on thin layer chromatography (TLC) plate. The plate was developed in (96% diethyl ether, 3% methanol and 1% water) and spot visualized at 365 um wavelength before scanning on TLC.

2.2 Experimental Design and Birds

Two hundred and forty one- week old Arbor Acre broilers used were randomly distributed into six dietary treatments using the simple completely randomised design. The birds were fed basal diet for one week before allotting them into their pens with four replicates for each treatments containing 10 birds per replicate. Normal routine management exercise was carried out and signs of disease and mortality were recorded.

2.3 Diets Preparation

Dietary treatments consisted of the basal diet (15 ppb aflatoxin) (diet formulated with normal maize, the aflatoxin in the diet was as a result of other ingredients added), other diets were formulated with rejected maize (the maize contain 32 ppb aflatoxin). These two diets contained either 2% honey or charcoal to form experimental other diets as shown in Table 1.

2.4 Sample Collection

Data on feed intake, final weight gain and body weight gain were collected during the experiments. At the end of the experiment, 54 birds were slaughtered for gut morphology (duodenum, jejunum and ileum) and histopathology on the organs.

2.5 Histopathology

Post-mortem examinations were performed on fifty four birds from all the treatments, three from each replicate. After being slaughtered, samples of liver, kidney and bursa of fabricius were collected and fixed in 10% neutral buffered formalin. Fixed tissues were trimmed, embedded in paraffin, sectioned at 4μm, and stained with hematoxylin and eosin stain. Tissue samples from all treatments groups were examined microscopically.

2.6 Statistical Analysis

Data were analysed using analysis of variance (ANOVA) of the General Linear Model procedure of SAS software, [21]. Differences among treatments means were tested for significance by using Multiple Range Duncan Test.

3. RESULTS AND DISCUSSION

3.1 Growth Performance

Table 2 shows performance characteristics of broiler birds fed aflatoxin contaminated feed supplemented with charcoal and honey as anti-aflatoxin additives. Although it has been believed that contamination of the diet with aflatoxin could detrimentally affect feed quality and thus bird performance, 0.032 ppm or 32 ppb aflatoxins in broiler diets did not impair broiler performance in this study. This lack of difference in growth performance might be due to low levels of AF (32 ppb) within a period of 42 days. This statement is in agreement with the report by Mahmoud et al. [22], who found that there were no changes in performance of broilers fed diet with lower level of AF (70 ppb AF) for a shorter period of 21 – 42 days of age. The reason for no changes in production parameters is because the birds were fed diet low in AF level (50 – 100 ppb AF) within a period of 46 days (Oguz et al. [20,21]; Ortatatli et al. [23]; Magnoli et al. [24]. Miazzo et al. [12] however reported that BW gains were lower (P < 0.05) for broilers that were fed AF in their diets.

Table 1. Gross composition of experimental diets (kg)

Composition (%)	CTL	AFL	CTL-Ch	AFL-Ch	CLT-Ho	AFL-Ho
Normal maize	60	-	60	-	60	-
Rejected maize	-	60	-	60	-	60
Groundnut cake	18	18	18	18	18	18
Soyabean meal	15	15	15	15	15	15
Fish meal	2	2	2	2	2	2
Dicalcium Phosphate	4	4	4	4	4	4
Vitamin-mineral Premix	0.25	0.25	0.25	0.25	0.25	0.25
Salt	0.25	0.25	0.25	0.25	0.25	0.25
Lysine	0.26	0.26	0.26	0.26	0.26	0.26
Methionine	0.24	0.24	0.24	0.24	0.24	0.24
Calculated composition values						
M. E (kcal/kg)	3013	3013	3013	3013	3013	3013
C.P (%)	23	23	23	23	23	23
Calcium (%)	1	1	1	1	1	1
Av. P (%)	0.98	0.98	0.98	0.98	0.98	0.98
Aflatoxin Quantified (ppb)	15	15	15	15	15	15

CTL= Normal diets (positive control with 15 ppb AF); AFL=Rejected maize diets (negative control with 32ppb Aflatoxin, AF); CTL-Ch=Positive control with 15 ppb AF plus 2% charcoal; AFL-Ch=Rejected maize diets with 32ppb Aflatoxin, AF plus 2% charcoal; CTL-Ho= Positive control with 15 ppb AF plus 2% honey; AFL-Ho= Rejected maize diet (32ppb AF) plus 2% honey M.E=Metabolisable energy, C.P.= Crude Protein, Av.P= Available Phosphorus

Table 2. Performance characteristics of broiler birds fed aflatoxin contaminated feed supplemented with charcoal and honey as anti-aflatoxin additives

Treatment	CTL	AFL	CTL-Ch	AFL-Ch	CTL-Ho	AFL-Ho	SEM	P-Value
ILW (g/bird)	173[b]	187[a]	178[ab]	178[ab]	158[c]	167[bc]	3.6	0.1
FLW (g/bird)	1475	1566	1533	1485	1481	1488	59.0	0.3
BWG (g/bird)	1297	1331	1369	1324	1323	1320	58.6	0.4
TFC (g/bird)	2591	2591	2709	2818	2637	2497	93.7	0.1
CFCR	2.7[a]	2.0[ab]	1.8[b]	2.1[ab]	2.2[ab]	1.9[ab]	0.2	0.2

abc means with different superscript on the same row are significantly different (P=.05)
ILW: initial live weight; FLW: final live weight; BWG: body weight gain; TFC: total feed consumed; CFCR: cumulative feed conversion ratio. CTL-normal diet (positive control i.e. diet containing 15ppb Aflatoxin), AFL-Rejected maize diet (negative control i.e. diet containing 32ppb Aflatoxin), CTL-Ch: normal diet (positive control)+2% charcoal, AFL-Ch: rejected maize diet(negative control) +2% charcoal, CTL-Ho: normal diet (positive control) + 2% honey, AFL-Ho: rejected maize (negative control) +2% honey

The lowered growth rate experienced upon feeding AF was due to reduction in body utilization of protein and energy Smith and Hamilton, [25]; Ianza et al. [26] vis-à-vis impaired nutrient absorption and reduced pancreatic enzymes for digestive purposes Osborne and Hamilton, [27] and subsequently appetite Sharline et al. [28]. There was an improvement in the feed conversion ratio of birds fed diets supplemented with the mycotoxin binder charcoal at 2% (AFL-Ch). This finding clearly supports previous reports of Murthy and Devegowda [6,7].

3.2 Intestinal Morphology

According to Bouhet et al. [29], the GIT is the first organ to come in contact with chemicals, natural toxins and foods and such should be affected with greater potency compared to other organs. Table 3 shows the results of intestinal morphology. There is a decrease in the villus height of birds duodenum and ileum but an increase in the jejunum (P<0.05) of birds fed rejected maize diets (negative control with 32 ppb Aflatoxin, AF). Cavret and Lecoeur [30] and Agence Française de Sécurité [31] explained that about >80% of aflatoxins are absorbed at the duodenum part of the intestines. As such mycotoxins always compromise intestinal epithelium either before or throughout the entire intestines by non-absorbed toxins. The ratio of villus height to crypt depth in the duodenum decreased but experienced increase in the jejunum and ileum section. However, the supplementation of 2% charcoal to the diets contaminated with 32 ppb AF i.e. AFL-Ch diet showed a significant (P<0.05) increase in

the villus height of duodenum and ileum. Although, 2% honey supplementation showed an increase in the villus height of birds duodenum no effects on jejunum and ileum. The ratio of villus height to crypt depth of the three intestinal segments follows the same trend. It is clear that increasing the villus height increased surface area vis-a-vis greater absorption of available nutrients. The villus height to crypt depth ratio according to Caspary [32] reflects differences in the digestion and absorption of the small intestines. Applegate et al. [33] reported that crypt depth of gut increases linearly with aflatoxin concentrations vis-à-vis influencing the villus crypt ratio.

A previous study by Girish and Smith [34] reported that grains naturally contaminated with DON significantly reduced the height, width, and surface of villus in the duodenum and jejunum of broilers. The present are in agreement with Yang et al. [35] statement that there was a decrease significantly in villus height and the ratio of villus height to crypt depth when broilers birds were fed daily with diets contaminated with AFB_1 and AFB_2. Long-term exposure of AFB_1 and AFB_2 mainly would affect the morphology of the duodenum as a result of stimulating proximal gastrointestinal tract Yang et al. [35]; the characteristics of the gut morphometric attributes vis-à-vis altering nutrients absorption. Girgris et al. [36] observed an increase in the villus height of jejunum and ileum of birds fed a contaminated diet with Fusarium mycotoxins which is contrary to the present result. The authors suggested a compensation for the reduced surface area of the duodenum villi resulting from reduced villi heights in these birds.

Table 3. Gut parameters of broiler birds fed aflatoxin contaminated feed supplemented with charcoal and honey as anti aflatoxin additives

Treatment	CTL	AFL	CTL-Ch	AFL-Ch	CTL-Ho	AFL-Ho	SEM	P-value
Duodenum	(mmX10^2)							
Villus height	270^{ab}	181^d	280^a	235^c	243^{bc}	214^c	9.9	0.4
Crypt depth	31^b	45^a	24^c	45^a	30^{bc}	45^a	2.4	0.6
Villus width	18^b	21^b	19^b	40^a	20^b	21^b	2.2	0.7
VH/CD	8^a	4^b	8^a	5^b	8^a	5^b	0.6	0.3
Ileum	(mmX10^2)							
Villus height	147^b	138^{bc}	174^a	182^a	118^c	116^c	7.8	0.3
Crypt depth	25^{ab}	19^b	26^a	20^b	22^{ab}	26^{ab}	2.0	0.1
Villus width	16^{ab}	13^b	17^a	16^{ab}	16^{ab}	16^{ab}	1.1	0.3
VH/CD	6^{bc}	7^b	6^{bc}	12^a	6^{bc}	5^c	0.7	0.5
Jejunum	(mmX10^2)							
Villus height	162^d	205^a	157^d	167^{cd}	184^{bc}	203^{ab}	5.9	0.4
Crypt depth	24^{bc}	25^{bc}	30^{ab}	28^{ab}	32^a	20^c	1.9	0.1
Villus width	19^b	18^b	18^b	17^b	25^a	17^b	0.7	0.5
VH/CD	6^b	8^a	5^b	6^b	6^b	7^{ab}	0.5	0.1

*abc means with different superscripts on the same column are significantly different (P=.05)
CTL-normal diet (positive control i.e. diet containing 15 ppb Aflatoxin), AFL-Rejected maize diet (negative control i.e. diet containing 32 ppb Aflatoxin), CTL-Ch: normal diet (positive control)+2% charcoal, AFL-Ch: rejected maize diet(negative control) +2% charcoal, CTL-Ho: normal diet (positive control) + 2% honey, AFL-Ho: rejected maize (negative control) +2% honey, VH/CD: Villus height to crypt depth ratio*

4. CONCLUSION

It can be concluded from the study that charcoal was able to prevent the absorption of the toxins into the enterocyte of the animal vis-à-vis giving better bird performance and gastrointestinal tracts attributes. However, honey was therapeutic in nature as it healed sours generated by mycotoxins absorption on the surface of the organs.

COMPETING INTERESTS

Authors have declared that no competing interests exist.

REFERENCES

1. Hussein HS, Brasel JM. Toxicity, metabolism, and impact of mycotoxins on humans and animals. Toxicology. 2001;167:101-134.
2. Wu F, Munkvold GP. Mycotoxins in ethanol coproducts: modeling economic impacts on the livestock industry and management strategies. Journal of Agricultural and Food Chemistry. 2008;56:3900–3911.
3. Zhang Y, Caupert J. Survey of mycotoxins in U.S. distillers dried grains with solubles from 2009 to 2011. Journal of Agricultural and Food Chemistry. 2012;60: 539–543.
4. Dutta TK, Das P. Isolation of aflatoxigenic strains of *Aspergillus* and detection of aflatoxin B1 from feeds in India. June, 2001, 0301-486X (Print) 1573-0832 (Online). 2001;151(1):39–44. [PubMed].
5. Sundu B, Hatta U, Damry HB. Comparison of Mycotoxin Binders in The Aflatoxin B1-Contaminated Broiler Diets. Proceeding of the 2nd International Seminar on Animal Industry | Jakarta. 2012;257-262.
6. Devegowda G, Murthy TNK Mycotoxins: Their effects in poultry and some practical solutions. In The Mycotoxin Blue Book; Diaz, D.E., Ed.; Nottingham University Press: Nottingham, UK. 2005;25–56.
7. Murthy TNK, Devegowda G.. Mycotoxins: their effects in poultry and some practical solutions. Ed. DE Diaz, Nottingham: Nottingham University Press. 2005;25-56.
8. Yiannikouris A, Jouany JP. Les mycotoxines dans les aliments des ruminants, leur devenir et leurs effets chez l'animal. INRA Productions animals; 2002.
9. Lopez-Garcia R, Park DL. Effectiveness of postharvest procedures in management of mycotoxin hazards. In: Mycotoxins in Agriculture and Food Safety (Sinha KK, Bhatnagar D, eds.), Marcel Dekker, Inc., New York, NY. 1998;407-433.
10. Sinha KK. Detoxification of mycotoxins and food safety. In: Mycotoxins in Agriculture and Food Safety. (Sinha KK, Bhatnagar D,

eds.), Marcel Dekker, Inc., New York. 1998;381-405.

11. Ibrahim IK, Shareef AM, Al-Joubory KMT. Ameliorative effects of sodium bentonite on phagocytosis and Newcastle disease antibody formation in broiler chickens during aflatoxicosis. Research in Veterinary Science. 2000;69:119–122.

12. Miazzo R, Rosa C, de Queiroz CE, Magnoli C, Chiacchiera S, Palacio G, et al. Efficacy of synthetic zeolite to reduce the toxicity of aflatoxin in broiler chicks. Poultry Science. 2000;79: 1–6.

13. Rosa CAR, Miazzo R, Magnoli C, Salvano M, Chiacchiera SM, Ferrero S, et al. Evaluation of the efficacy of bentonite from the south of Argentina to ameliorate the toxic effects of aflatoxin in broilers. Poultry Science. 2001;80:139-144.

14. Pasha TN, Farooq MU, Khattak FM, Jabbar MA, Khan AD. Effectiveness of sodium bentonite and two commercial products as aflatoxin absorbents in diets for broiler chickens. Animal Feed Science and Technology. 2007;132:103-110.

15. Oguz H, Kurtoglu V. Effect of clinoptilolite on performance of broiler chickens during experimental aflatoxicosis. British Poultry Science. 2000;41:512–517.

16. Oguz H, Kurtoglu V, Coskun B. Preventive efficacy of clinoptilolite in broiling during chronic aflatoxin (50 and 100 ppb) exposure. Research in Veterinary Science. 2000;69:197-201.

17. Chestnut AB, Anderson PD, Cochran MA, Fribourg HA, Twinn KD. Effects of Hydrated Sodium Calcium Aluminosilicate (HSCAS) on Fesue Toxicosis and Mineral Absorption. Journal of Animal Science. 1992;70:2838 -2846.

18. Whitlow LW. Evaluation of mycotoxin binders. Proceedings of the 4th mid-atlantic nutrition conference March. 2006;29-30,

19. Edrington TS, Kubena LF, Harvey RB, Rottinghaus GE. Influence of superactivated charcoal on the toxic effects of aflatoxin or T-2 toxin in growing broiler. Poultry Science. 1997;76:1205-1211.

20. Wellford TE, Eadie TT, Llewellyn GC. Evaluating the inhibitory action of honey on fungal growth, sporulatiion and aflatoxin production. Z. Lebensm. Unters. Forsch. 1978;166:280 – 283.

21. SAS. SAS User's Guide statistics. SAS Inc. Cary, North Carolina, edition; 2009.

22. Mahmood T, Pasha TN, Khattak FM. Comparative Evaluation of Different Techniques for Aflatoxin Detoxification in Poultry Feed and Its Effect on Broiler Performance, Aflatoxins – Detection, measurement and control. Dr Ireneo torres-pacheco (ed,); 2011. ISBN. 978-953-307-711-6.

23. Ortatatli M, Oguz H, Karaman M. Evaluation of pathological changes in broilers during chronic aflatoxin (50 and 100 ppb) and clinoptilolite exposure. Research in Veterinary Science. 2005;78:61–68.

24. Magnoli AP, Monge MP, Miazzo RD, Cavaglieri LR, Magnoli CE, Merkis CI, et al. Effect of low levels of aflatoxin B on performance, biochemical parameters, and aflatoxin B in broiler liver tissues in the presence of monensin and sodium bentonite. Poultry Science. 2011;90:48–58.

25. Smith JW, HamiltonPB. Aflatoxicosis in broiler chickens. Poultry Science. 1970;49:207-215.

26. Lanza GM, Washburn KW, Wyatt RD. Variation with age in response of broilers to aflatoxin. Poult. Sci. 1980;59:282–288.

27. Osborne DJ, Hamilton PB. Decreased pancreatic digestive enzymes during aflatoxicosis. Poult. Sci. 1981;60:1818–1821.

28. Sharline KSB, Howarth BJ, Wyatt RD. Effect of dietary aflatoxin on reproductive chicks. Poultry Science. 1980;72:651-657.

29. Bouhet S, Hourcade E, Loiseau N, Fikry A, Martinez S, Roselli M, et al. The mycotoxin fumonisin B1 alters the proliferation and the barrier function of porcine intestinal epithelial cells. Toxicological Science. 2004;77:205-211.

30. Cavret S, Lecoeur S. Fusariotoxin transfer in animal. Food Chemistry Toxicology. 2006;44:444–453.

31. French Agency for Food Safety Risk Assessment. to the presence of mycotoxins in food and feed chains; French Agency for Food Sécurité Sanitaire Maisons- Alfort, France. 2009;1–308.

32. Caspary WF. Physiology and pathophysiology of intestinal absorption. America Journal of Clinical Nutrition. 1992;55:299-308.

33. Applegate TJ, Schatzmayr G, Pricket K, Troche C, Jiang Z. Effect of aflatoxin culture on intestinal function and nutrient loss in laying hens. Poultry. Science. 2009;88:1235–1241.

34. Girish CK, Smith TK. Effects of feeding blends of grains naturally contaminated with *Fusarium* mycotoxins on small intestinal morphology of turkeys. Poultry Science. 2008;87:1075–1082.

35. Yang J, Bai F, Zhang K, Bai S, Peng X, Ding X et al. Effects of feeding corn naturally contaminated with aflatoxin B1 and B2 on hepatic functions of broilers. Poultry Science. 2012;91:2792-2801.

36. Girgis GN, Barta JR, Brash M, Smith TK Morphologic changes in the intestine of broiler breeder pullets fed diets naturally contaminated with *Fusarium* mycotoxins with or without coccidial challenge. Avian Disease. 2010;54:67–73.

3

Effect of Natural Mating Frequency and Artificial Insemination on Fertility in Rabbits and Their Cyto-genetic Profile (X-chromatin)

P. K. Ajuogu[1*], U. Herbert[2] and M. A. Yahaya[3]

[1]Department of Animal Science and Fisheries, Faculty of Agriculture, University of Port Harcourt, P.M.B.5323 Choba, Port Harcourt, Nigeria.
[2]Department of Animal Science, Micheal Okpara Federal University of Agriculture Umudike, Abia State, Nigeria.
[3]Department of Animal Science, Faculty of Agriculture, Rivers State University of Science and Technology, Nkpolu Orowurukwu Diobu, Port Harcourt, Rivers State, Nigeria.

Authors' contributions

This work was carried out in collaboration between the authors above. Author PKA designed the study, collected raw data from the field, performed the statistical analysis, wrote the protocol, and wrote the first draft of the manuscript. Authors UH and MAY managed the analyses of the study. Author MAY managed the literature searches. All authors read and approved the final manuscript.

Editor(s):
(1) Javier Alarcon Lopez, School of Engineering, University of Almería La Cañada de San Urbano, Spain.
Reviewers:
(1) Maria Grazia Cappai, Department of Agricultural Sciences, University of Sassari, Italy.
(2) Anonymous, Nigeria.
(3) Anonymous, India.

ABSTRACT

A study was conducted to determine the cyto-genetic profile of breeding rabbits and the impact of natural mating frequency and artificial insemination on its fertility status. Twenty four (post-pubertal) does aged 8-9 months and four matured (aged 10-12 months) fertile bucks of New Zealand white breed, were randomly assigned to four experimental groups (A, B, C and D) in a Completely Randomized Design (CRD). Groups A to C had been subjected to different level of daily mating of once a day (treatment A- the control), twice daily (treatment B) and thrice daily (treatment C) mating pattern. Animals of group D were subjected to artificial insemination. The experiment lasted for six months, during which two parities were obtained. The mean conception rate, kindling rate

Corresponding author: Email: peter.ajuogu@uniport.edu.ng;

24

and litter size showed significant differences (p<0.001) among experimental groups according to increasing level of frequency of daily mating. Significant difference (p<0.05) was also observed on litter weight when decreasing the order of daily mating frequency. The frequency of daily mating had no influence (p>0.05) on gestation length. The X-chromatin incidence of does with zero conception and their males were within a normal range of 2-12% and 0-2%, respectively. It can be concluded that daily mating frequency has a favorable influence on reproductive parameters examined in this study.

Keywords: Breeding pattern; X-chromatin; rabbit doe; artificial mating; fertility.

1. INTRODUCTION

Reproduction is a complex biological phenomenon which occurs in all living organisms. Its significance and manipulation are the secret tools for the improvement of animal fertility [1,2]. Rabbit's sexual behavior and breeding potential are all influenced by a wide range of external and internal stimuli [3,4].

Female rabbits are reputed to have complex and peculiar reproductive pattern. Apart from having high incidence of pseudo-pregnancy, the female rabbits are also considered poly-estrous, or as having no cyclic or regular estrus [3,4]. These biological features seems to presuppose that the non-pregnant female rabbit would accept her male counterpart for mating at every presentation and indeed conceive at each mating since it is an induced ovulator. [3,5] reported that other estrous signs in female rabbits even though less visible, than other farm species undergoes rhythmic weekly (5-7 days) estrous cycle. Influence of mating frequency is important in the reproductive performance of farm animals including rabbits [6,4]. [7] reported a 0.28 increase in ovulation rates of the mated groups of sows. [8] compared fertile results of sows inseminated three times in each estrous period with those sows inseminated twice. He reported both an increased conception rate and an increased litter size. [9] Reported that clitoral stimulation following artificial insemination of beef cow hastened the onset of ovulation with cervical stimulation reducing the time from the beginning of estrous to the luteinizing hormone (LH) surge suggesting that a neutral path way may exist between the genital system and the hypothalamic pituitary axis.

The use of X-chromatin (drumstick) in diagnosing major X-related reproductive irregularities has contributed greatly in the detection of kind of reproductive anomalies and stunted growth in the animal industry [10,11]. The technique could thus be used in culling individuals exhibiting numerical or morphological abnormalities of X-chromatin appendages early in life so as to reduce the cost of production [12,3]. In modern genetic parlance, X-chromatin evaluation refers to the analysis of the X-chromosome only, without reference to the Y chromosome. The X-chromosome has successfully been used in domestic animals to predict the cytogenetic or genetic merit of various economically important species. These include early detection of sex chromosomal and developmental anomalies which considerably impair fertility, prediction of the growth potential of neonates [13,14]. The principle is that there are as many Barr bodies (i.e. X-chromatin appendages) for the existence or presence of two X-chromosomes in a normal female animal cell. This implies that the normal fertile animal is expected to show at least one X-chromatin appendage in a preponderant proportion ie, in at least 90% of its cells [10]. The objectives of this research are: (1) to establish or ascertain normal fertility in female rabbits through X-chromatin evaluations (2) to evaluate the influence of natural mating frequency on the reproductive performance of female rabbits (3) to compare the influence of natural mating and artificial insemination on conception rate, gestation length, litter size, litter weight, survivability and weaning weight/age in rabbits.

2. MATERIALS AND METHODS

Twenty four post pubertal female rabbits, aged 8-9 months and four mature fertile bucks aged 10-12months of New Zealand white breed were randomly assigned to four experimental groups A,B,C and D in a Completely Randomized Experimental design (CRD). Each experimental group was further replicated twice with three female rabbits per replicate. The treatments were properly tagged for easy identifications indicating the breeding pattern and timing, as detailed below:

A = Morning mating 8:00-9:00 am, (once mating per day) (control),

B = Morning mating 8:00-9:00 am, afternoon mating 1:00-2:00 pm (twice mating per day),

C = Morning mating 8:00-9:00 am, afternoon mating 1:00-2:00 pm and evening mating 7:00- 8:00 am (thrice mating per day),

D = Artificial insemination (i.e insemination by artificial or mechanical means).

For collection of rabbit semen, a reinforced polyvinylchloride (PVC) tube (inner diameter 2.8 cm, outer diameter 3.8 cm) was cut to a length of 3.0 cm and referred to as the lager tube (LT). A rigid plumbing hose (inner diameter 1.7 cm and outer diameter 2.7 cm) was also cut to the same length and called the smaller tube (ST). These were used to construct artificial vagina (AV) according to [15]. The artificial vagina (AV) unit was warmed by putting it in warm water at 60°C for 10-15 minutes in a rubber container according to [15]. A mature non gravid doe was used as the teaser. As the buck mounts and makes a thrust on the teaser doe, the pre-warmed AV was quickly applied from side of the erect penis prior to intromission of the penis in the AV, which elicited ejaculation within few seconds. The semen volume (in ml) was read directly from the collection tube, and used for the insemination.

The female was taken to the male for service and returned thereafter. The males belonging to each experimental group were used to serve the females in that group. Females of groups A-C were naturally mated with their respective males while females of group D were artificially inseminated by the semen collected from the male of that group.

For the artificial insemination procedure, the female is placed into a restraining box, after this, the tail was lifted, and an insemination pipette with a bend approximately 8 cm from the end, and inserted into the vagina at an angle of 45° in order progress beyond the pelvic rim according to [5]. Semen is deposited intra-vaginally.

Palpation for pregnancy was carried out on day 14 in accordance with the procedure described by [6]. Does must be relax and sitting naturally whereby fingers were gently run along the abdomen between the back legs, small bead-like lumps would be felt if the doe was pregnant. Blood samples were collected from the females that could not conceive after mating. Sterile syringes and hypodermic needles were used to collect blood from their ear veins. After collection, blood samples were immediately decanted into a

sterile heparinized sample tubes to avoid coagulation and immediately taken to the laboratory for X- chromatin evaluation. A drop of the whole blood was dropped on a glass slide and a smear made and allowed to dry. After that, samples were fixed in 90% ethanol for 3 minutes and stained after drying for 5 minutes with Leishman's stain, then rinsed with tap water, and left to dry. Finally, blood slides were examined under oil immersion (x 100) for visualization of Barr bodies. The males belonging to the treatment groups that could not conceive were also screened for X-chromatin status.

Two hundred polyrnorphonuclear neutrolear (PMN) per slide were counted in order to determine the percentage incidence of drumstick, as follows:

% incidence =

$$\frac{\text{Number of drumsticks observed}}{\text{Number of PMN observed}} \times \frac{100}{1}$$

Variables used to measure reproductive parameters were conception rate, gestation length, litter size, litter weight, pseudo-pregnancy, weaning weight and mortality. Two parities were obtained. Data of reproductive parameters were collected and subjected to analysis of variance according to [16] and their means separated by Duncan multiple range test by [17].

3. RESULTS AND DISCUSSION

The results of the effect of mating frequency and artificial insemination on reproductive parameters in both parities are summarized in Table 1. The result showed that conception rate was significantly influenced (P<0.05) by natural mating frequency in parity 1 and 2 than the group handled with artificial insemination. For first parity group, treatments C and D recorded the highest and the lowest conception rates with values of 83.3% and 16.7%, respectively Table 1. While in parity two, the highest and lowest conceptions rates were found in groups C and A (100% and 33.3%, respectively). The average gestation length had no significant (p<0.05) effect among experimental groups on both parties. Average kindling rate showed a significant increase (p<0.05), except for artificially inseminated group (33.3%, 66.7%, 91.7% and 8.4% for groups A, B, C, and D, respectively. Also, the results of the average litter size and litter weight showed significant difference among groups (p<0.05).

Litter size progressively increased according to the mating frequency in both parity (parity 1: 3, 0, 4.67, 5.8, 4.0; and parity 2: 3.0, 4.25, 6.17, 0.0 for treatments A, B, C, and D, respectively). But the litter weight was negatively affected by the mating pattern, the higher the frequency of mating, the lower the litter weight. Pseudo-pregnancy was also observed among the groups in both parities. The highest value was recorded in treatment D in both parities. Mortality of does and kids was also observed.

The results of X-chromatin (Barr body) incidence are presented in Table 2. The average drumstick incidence in female rabbits was 4.45%, while the average values in males was 0%. It was found no X-chromatin incidence in one of the females (F6) of group A.

Fairly high conception rates ranging from 33.3-100% were observed when increasing the frequency of mating in this study. This finding support that of [18], who reported increased reproductive performance, conception rate and litter size of rabbits inseminated more than once.

However, inseminated group recorded the lowest conception rates during the first parity (16.7%) and the second parity (0%). This observation may be related to the insufficient physiological stimulation of the females [19,20] resulting in low conception rate or non-conception. This supports the assertion of that rabbits are induced ovulators [1,18]. This fact implies that endocrinal gland could not receive enough stimuli (signal for concomitant release of ovulation hormone; luteinizing hormone and follicle stimulating hormone [19] for induction of ovulation, it is possible therefore, that the ovum or ova were not released for fertilization by the semen introduced through artificial insemination, hence the poor or no conception observed. Also [4] reported that frequency of mating either natural or artificial, improves some reproductive parameters (conception rate, litter size and kindly rate). However, the reports of [18,5,1] revealed that rabbit undergo a rhythmic weekly (5-7 days) estrus cycle. This may be the reason for low conception rate (16.7%) recorded in treatment D among the artificially bred does. The insemination may have been co-incidental on their estrus period since there was no tactile or physiological stimulation. The null conception recorded for the same does (group D) in the second parity supports this assertion. [20] similarly observed poor conception rate in

artificially inseminated rabbits. He implicated lack of ovulation induction as being responsible.

However, since rabbits are induced ovulators [21] other than natural mating, stimuli, including direct vaginal stimulation with a vaginal swab or a glass rod [22] and administration of hCG [23,24] or GnRH [25] could have helped to induced ovulation. Also [26,27,28] have reported improvement of reproductive performance by the use of methods enabling the induction and synchronization of oestrus. Also [21], reported that hormonal treatments have been widely used in recent years, especially the use of prostaglandin F2 alpha (PGF2α) or its analogues to improve doe receptivity and improving reproductive parameters. The poor reproductive efficiency, observed in group D (artificially bred group), confirm the fact insemination should have preceded induction

The result showed that mating frequency and artificial insemination has no effect on gestation length ($p<0.05$). This result agrees with [29,30] who reported a mean gestation length of 31.6 days in a well-managed rabbit breeding outfit. The natural mating frequency has significant influence ($P<0.05$) on kindling rate, which agrees with the assertion of [31] who implicated vaginal uterine stimuli (tactile stimulation through coitus) resulting in higher conception rate, kindling rate and greeter viability in-utero.

Also, litter size showed significant differences ($p<0.05$) among experimental groups, with the highest values recorded in groups B (5,4) and C (6,6). This fact could probably be due to increase in neural and endocrinal release of luteinizing hormone (LH) and follicle stimulating hormone (FSH) during copulation, which act on the ovary causing more ova release. This agrees with [32] who observed that copulation in rabbits stimulates the immediate release of luteinizing hormone with peak serum levels reached in 1-2 hours post copulation. This implies that the more the number of eggs released due to coital stimulation, the more the available sperm cells in the reproductive system will fertilize them. This obviously will affect the number of litters and other reproductive parameters (conception rate, litter size and kindling rate).

The mean percent drumstick incidence obtained from animals of experimental groups that had no conception ranged from 2.5 to 15.0%, with a mean value of 4.5%. This observation agrees

Table 1. Mean values of reproductive parameters of New Zealand white does in the first and second parity

Reproductive Parameters	Parity 1 Experimental groups (mean + SEM)				Parity 2 Experimental groups (mean + SEM)			
	A	B	C	D	A	B	C	D
Conception rate (%)	33.3±0.0^c	66.7±0.0^a	83.3±0.0^a	16.7±0.0^d	33.3±0.0^c	66.7±0.0^b	100.0±0.0^a	-
Gestation length(days)	30.5±1.0	31.3±0.0	30.6±00	1.0±0.0	31.5±16	31±1.0	31.0±0.0	-
Kindling rate (%)	33.3±0.0^a	66.7±0.0^d	91.7±1.0^c	8.4±0.0^d	-	-	-	-
Litter size (Number)	3.0±0.0^a	4.7±1.0^c	5.8±1.0^c	4.0±0.0^b	3.0±1.0^a	4.3±0.0^{ab}	6.2±0.0^c	-
Litter weight (g)	85.2±1.9^a	72.8±2.4^b	68.0±2.4^b	5.7±0.0^{ab}	81.7±57^a	71.3±31^{ab}	67.4±2.3^b	-
N° of pseudo-pregnancy	2	2	0	-	-	-	3	-
N° of still births	-	-	-	-	-	-	-	-

Values in the same line with different lower scripts show significant differences among experimental groups (p<0.05)

with that of [33] who documented that some mammals such as humans and cattle showed an X-chromatin incidence of 2% - 12% in a normal female. Also [34] reported an X-chromatin incidence of 0.5% per 200 polymorpho-nuclear neutrophils (pmn). These observations therefore suggest that the females in treatments D (F1 -5), A (F6-9) and B (F1O-11), having no conception Table 2 may not have been affected by any form of health or management problems. According to this study, the experimental treatment was implicated in the observed variations. Therefore, it could be implied that the influence of mating once, (treatment A) twice (treatment B) and artificial mating without artificial induction (treatment D) may have been the possible cause of the null conception observed within the groups in reference. Nevertheless, the zero percent (0.0%) X-chromatin (drumstick)' incidence of one of the females in treatment A (F6) may suggest the existence of a genetic (sex-chromosomal) disorder with the consequence of reproductive failure [35,36,37,11,14].

Furthermore, the values of X-chromatin in the males (0% drumstick) was in line with [34] who reported an X-chromatin appendages of 0.0% - 2.0% for normal mammalian males. The above observation therefore, suggests that genetic reproductive abnormalities may not have coursed the observed differences in reproductive and productive performances.

Table 2. X-chromatin status of the animals used in the present study

Animals	N° of Neutrophils	% incidence of drumstick
A1	200	3.60
A2	200	6.67
A3	200	15.40
A4	200	2.04
B5	200	2.04
B6	200	0.00
B7	200	4.17
F8	200	7.14
F9	200	4.55
F10	200	5.13
F11	200	2.50
F1	200	0.00
F2	200	0.00
F3	200	0.00
Mean	200	4.44

4. CONCLUSION

It was demonstrated that increased natural mating frequency has better influence on reproductive parameters (conception rate, litter size and kindling rate) than artificial insemination in rabbits.

Pseudo-pregnancy, one of the major constraints in effective reproduction in rabbits, can be reduced greatly by increasing frequency of natural mating.

The influence of artificial insemination on reproductive performance of rabbits may also be enhanced when the doe is properly induced physiologically (tactile induction).

COMPETING INTERESTS

Authors have declared that no competing interests exist.

REFERENCES

1. Berepubo NA, Nodu MB, Monsi A, Amadi EN, Reproductive response of pre-pubertal female rabbit to male presence and photoperiod. Int. J Anim Sci. 1993; 8:57-61

2. Dalton DC. An Introduction to practical animal breeding. Granda Publishing Limited St. Abband and London. 1987;12-16.

3. Berepubo NA, Umanah AA. Preliminary investigations into primary anestrous/ Infertility in a female rabbit (Dos). Int. J. Anim Sci. 1996;11:205-208.

4. Ajuogu PK, Ajayi FO. Breeding responses of New Zealand white does to artificial insemination under humid tropical environment. Animal Production Research Advances. 2010;6(1):46-48.

5. Fielding D. Rabbit: The Tropical Agriculturalist CTA/Macmillan Education Ltd. London. 1991; 21-25.

6. Ajuogu PK. Influence of mating frequency and artificial insemination on fertility in rabbit. M.Sc. Thesis, Rivers State University of Science and Technology. 2002;1–26.

7. Pitkjanen IG. Ovulation, fertilization and first Stages of embryonic development in Pigs. Izv. Akad. Nauk SSR. Ser Blot. 1955; 3:120-131.

8. Reed HCB. Artificial Insemination. In Control of Pig Reproduction ed. DJ.A. Cole and G.R. Fox Croft. Butteriorth, London. 1982;65-90.

9. Randel RD, Short RE, Christensen DS, Below RA. Effect of various Mat ing Stimuli on the LH Surge and Ovulation time Following Synchronization of Estrus in the Bovine. J. Anim. Sci. 1973;37:128-130.

10. Bhatia S, Shanker VV. Sex-chromatin test in the bovine: A cytodiagnostic tool for detection of poor breeders at birth. Proc. 10th Int. Congr. Anim. Reprod. 1984;10-14.

11. Berepubo NA, Pinherio LE, Basrur PK. Biological significance of X-chromosomes inactivation pattern in sub fertile cows carrying an X- autosome translocation. Discov. Innov. 1993b;5(1):57-62.

12. Omeje SI, Berepubo NA Nwankwo PC. X-chromatin study of native breeds of cattle and small ruminants in Nigeria. International J. Anim. Sci. 1994;9:181-184.

13. Satbir K, Vasudha S, Anju S, Parminder K. Buccal. Mucosal X-chromatin frequency in breast and cervix cancer. Anthropologist. 2006;8(4):223-225.

14. Wekhe SN, X-chromatin evaluation of reproductively abnormal ruminants, pigs, rabbits and man in Rivers State of Nigeria. PhD thesis. Rivers State University of Science and Technology, Nkpolu. Port Harcourt. 1998;123.

15. Herbert U, Adejumo DO. Construction and evaluation of an Artificial vagina for collection of rabbit semen. Delta Agric. 1996;2:99-108.

16. Steel RGD, IH Torrie. Principle and procedures of statistics. 2nd Ed. McGraw Hill Singapore; 1981.

17. Duncan DB. Multiple Range and Multiple F-test Biometric ewe. Acta Vet. Scand. 1955;16:145.

18. Paufler S. A compendium of rabbit production appropriate for conditions in developing country. Eschborn, Germany. 1985;115-130.

19. Murdoch WJ, Dunn TG. Alterations of in follicular stand hormones during the pre ovulation period in the Ewe. Bisit. Reprod. 1982;27:300.

20. Rahajo YC, Cheekke PR, Patton NM. Growth and reproductive performance of rabbits on a moderately low grude protein diet with or without Methionine or Urea Supplementation. J. Anim. Sci. 1986; 63:795-803.

21. Abdel-Azeem AS, Abdel-Azim AM, Hanan A. Hassan, Mona S. Ragab. Effects of prostaglandin F 2 alpha injection and rabbit does age on reproductive, hematological parameters and level of progesterone. Egypt. Poult. Sci. 2012;32(1):189-199.

22. Greulich WW. Artificially induced ovulation in the cat (Felis domestica). The Anatomical Record. 1934;58:217-224.

23. Wildt DE, Kinney GM, Seager SWJ. Gonadotropin induced reproductive cyclicity in the domestic cat. Laboratory Animal Science. 1978;28:301-307.

24. Cline EM, Jennings LL, Sojka NJ. Breeding laboratory cats during artificially induced estrus. Laboratory Animal Science. 1980; 30:1003-1005

25. Chakraborty PK, Wildt DE, Seager SWJ. Serum luteinizing hormone and ovulatory response to luteinizing hormone releasing hormone in the estrous and anestrous cat. Laboratory Animal Science. 1979;29:338-344.

26. Boiti C, Besenfelder U, Brecchia G, Theau-Clément M, Zerani M. Reproductive physiopathology of the rabbit doe. In: Recent Advances in Rabbit Sciences (Maertens, L. and Coudert, P. Editors). 2006;ILVO:3-19.

27. Theau-Clement M. Facteurs de réussite de l'insémination chez la lapine et méthodes d'induction de l'oestrus. INRA Prod. Anim. 2008;21:221–230. (In French).

28. Gogol P. Effect of prostaglandin F2α on reproductive performance in rabbit does. Ann. Anim. Sci. 2009;9:395–400.

29. Omole TA, Sonaiya FO. Nutritional Report International. 1991;23(4):729-739.

30. Aduku A0, Olukosi BO. Rabbit Management in the tropics code living book series, Au publications Abuja Nigeria; 1990.

31. Drugociu G. The importance of internal and external stimuli for sexual function in Sows. Zerg Prod. Postep Nauh Rolu. 1966;61:107-116.

32. Checke, PR Patton NM, Lukefohr SD, Mcnitt JI. Rabbit production. The interstate printers and Publishers, Illinois; 1987.

33. Basrur PK. Test for nuclear dimorphism. In: A guide lo genetic counseling in veterinary medicine, P.K, Basrur ed Ontario veterinary college. University of Guelph, Canada. 1984;92.

34. Davidson WM, Smith MD. A morphological sex difference in the Polymorph nuclear neutrophil. Brit. Med. J. 1954;2:67.

35. Long SE, Williams CV. Frequency of chromosmal abnormalities in early embryos of the domestic sheep (*Ovis aries*). J. Reprod Fert. 1980;58:197.

36. Hare WCD, Sigh EL, Betteridge KJ, Eaglleson MD, Randall GCB, Mitchell D, Bilton TJ, Trounson AO. Chromosome analysis of 159 bovine embryos colleted 12 to 18 days after Oestrus. Can. J. Genet. Cytol. 1980;22:615-262.

37. Bhatia S, Shanker V. Practical applications of sex chromatin studies in the investigation of distributed fertility in female cattle. Proceeding of the second World Congress on Genetics Applied to Livestock Production held on October 1982 at Madrid Spain. 1982;483.

4

Utilization of Dicalcium Phosphate and Different Bone Meals as Dietary Phosphorus Supplement in the Diets of *Clarias gariepinus* Fingerlings

I. A. Adebayo[1*] and G. H. Akinwumi[1]

[1]*Department of Forestry, Wildlife and Fisheries Management, Faculty of Agricultural Sciences, Ekiti State University, Ado Ekiti, Ekiti State, Nigeria.*

Authors' contributions

This work was carried out in collaboration between both authors. The two authors IAA and GHA designed and participated in the research work. Author IAA supplied the glass aquaria and bought the ingredients used for the experiment. The feed formulation and fish procurement at the commencement of the experiment were done by author IAA. Feed administration to fish for 70 days and weighing of fish were carried out by author GHA. The statistical analysis of data generated from the experiment was jointly done by the two authors. The manuscript was written by author IAA, while the literature searches was done by author GHA. The two authors read the final manuscript and approved its submission for publication.

Editor(s):
(1) Md. Yeamin Hossain, Department of Fisheries, Faculty of Agriculture, University of Rajshahi, Bangladesh.
Reviewers:
(1) Anonymous, Nigeria.
(2) Anonymous, India.
(3) Anonymous, Nigeria.
(4) Anonymous, Thailand.
(5) Anonymous, Brazil.

ABSTRACT

Phosphorus is most limiting mineral in the diet of fish and must be supplied in the right quality and quantity to prevent its deficiency or toxicity in fish. A 70 day feeding trial was conducted to evaluate the utilization of Dicalcium phosphate and different bone meals as phosphorus supplement in the diets of *Clarias gariepinus* fingerlings. At the start of the experiment, ten glass aquaria of size 70cm x 45 cm x 40 cm/each, filled with Well water up to 70L of its volume were stocked with one hundred (100) fingerlings (mean weight 6.00±0.02) g/one at 20 fish per treatment replicated twice using a complete randomized design. Five experimental diets (D_1-D_5) were formulated to be isocaloric

---

Corresponding author: Email: dayadeisrael@yahoo.com;

(11.1kcal/kg) and isonitrogeneous (40%Crude Protein). D_1 (control) was without Phosphorus (P) supplementation and P (0.44%) deficient, while D_2-D_5 were supplemented with Dicalcium phosphate (DCP), Chicken bone meal (CHBM), Clarias bone meal (CLBM) and Cattle bone meal (CABM) at 1.46%, 1.43%, 1.46% and 1.54% respectively based on the available P in each supplement to give 0.8% available P in the diets. The results showed that fish fed with D_3 (CHBM) was significantly ($p<0.05$) had the best growth performance in terms of Final Mean Weight Gain (FMWG) 14.40±0.14 g; Specific Growth Rate (SGR) 1.75±0.05 and Feed Conversion Ratio (FCR) 1.15±0.01. Phosphorus in the carcass of fish after the experiment was significantly ($p<0.05$) highest in D_5 (26.05±0.21)mg/g and least in fish fed D_1 (12.98±0.20) mg/g. The feeding trial established the necessity for phosphorus supplementation in the diets of *Clarias gariepinus* fingerlings for better growth and body mineralization.

Keywords: Dicalcium phosphate; bone meals; phosphorus; diets; growth and Clarias gariepinus.

1. INTRODUCTION

Quantitatively, 40 nutrients have been identified as necessary for the normal metabolic function of fish [1]. Minerals, especially phosphorus are important for optimal growth of cultivated fish. It guards against diseases such as lordosis (deformed backs and heads) as a result of abnormal calcification of bone [2]. Phosphorus is needed to help balance vitamins and other minerals which include vitamin D, iodine, magnesium and zinc. It is a vital component of organic phosphate such as adenosine triphosphate phospholipids, co-enzymes and DNA, which have major roles in metabolism of carbohydrates, fats and amino acids [3]. Phosphorus is essential for many intracellular processes notably; glycolysis, membrane maintenance, oxygen transport, muscle contraction and protein from oxidative damage [4].

Phosphorus must be provided in fish feed because of its low concentration in water [5]. According to [1] quantity of phosphorus that are available in water to fish are lower than 0.1ppm, as a result, the soluble phosphorus available to fish is less than 1% of what the fish actually needed to survive and for this reason phosphorus must be supply in the feed.

It has been demonstrated that phosphorus requirements are species-specific [5]. An aquaculturist must know the nutritional requirement of fish before compounding a feed. This information will help a farmer in formulating a feed that is nutritionally balanced for such fish particularly when they are cultured under intensive system [6]. However, bioavailability of dietary phosphorus is influenced by the digestibility of diet, particle size, and interaction with other nutrients, feed processing and water chemistry [7]. The optimal amount of phosphorus

supplementation in commercial feed is important economically and environmentally and must carefully balance to prevent signs of deficiency and minimize the urinary and fecal discharge of phosphorus into natural water which causes deterioration of water quality [7].

Little information is available on the phosphorus availability and utilization of bone meal prepared separately from individual animal bone species. Historically, bone meal is prepared from mixture of bone meal from undefined sources [8]. The evaluation of bone meal phosphorus and utilization is important in order to ensure that fish are not deficient or have marginal levels of phosphorus under the assumption that bone meal phosphorus is available completely [9]. There are variations in the availability and subsequently utilization of Phosphorus from different bone meals. This study therefore investigates the utilization of available phosphorus in Dicalcium phosphate, chicken bone meal, Clarias bone meal and Cattle bone meal as P supplement in the diets of *Clarias gariepinus* fingerlings.

2. MATERIALS AND METHODS

2.1 Experimental Procedure

The study was conducted from February 1st to 12th April, 2013 at the Faculty of Agricultural Sciences Laboratory, Ekiti State University, Ado-Ekiti, Nigeria. Prior to the start of the experiment, one hundred and twenty (120) fingerlings of *Clarias gariepinus* were purchased from Success Fish Breeding and Poultry Farms Nig. Ltd., Akure, Ondo State, Nigeria. Fish were temporarily kept in already prepared indoor concrete tank of size 4 m × 4 m × 1.5 m for a week and fed with commercial catfish feed (Coppens) of size 2 mm and 40% Crude Protein to acclimatize them.

At the start of the experiment, ten glass aquaria tanks of size 70 cm × 45 cm × 40 cm/one were filled with clean Well water to 70 litres of their capacity. Fish were then counted (10/tank) and weighed (mean weight 6.00±0.02) g/one, using Digital Pocket Scale (Model: 1000 g/0.1) and randomly stocked in a complete randomized design. The five treatments, replicated twice according to the test diets including the control were fed for a period of 70 days.

2.2 Experimental Diets and Feeding Procedure

All the ingredients used in feed preparation were purchased from Metrovet Feed mill in Ado-Ekiti, Ekiti State, Nigeria. Ingredients were analyzed for proximate composition prior to feed formulation according to [10]. The gross composition of the experimental diets was presented in Table 1. Experimental diets contained different proportion of feed ingredients to achieve different Phosphorus (P) inclusion levels. The control diet (D_1) was P (0.44%) deficient (without supplementation), while D_2- D_5 were supplemented with Dicalcium phosphate (DP) $_{1.46\%}$; Chicken bone meal (CHBM) $_{1.43\%}$; Clarias bone meal (CLBM)$_{1.46\%}$ and Catlle bone meal (CABM)$_{1.54\%}$ to give 0.8% P level in the test diets according to Adebayo (2011). In preparing the diets, dry ingredients including the bone meals (plate 2-4) were ground to a powdery form in a willey mill to enhance optimum utilization and digestibility. Plate 1 is dicalcium phosphate

(DCP) already in powdery form and there was no need for further milling before adding it to other ingredients to formulate D_2. Diets were thoroughly mixed with cod liver oil and pelleted using Hobart A 200 pelleting machine with a 2.0 mm die. They were immediately sun dried and packed in a labeled tight container and kept in the refrigerator prior to use. Isonitrogenous and isocaloric diets (40% CP) as recommended by Faturoti et al. [11] were used during the experimental study. Fish were fed to satiation twice daily at 09:00 and 16:00 for 70 days while the weights of the experimental fish were measured bi-weekly to calculate their response to feed.

2.3 Water Quality Parameters

Water quality parameters such as pH, DO, temperature and nitrate were closely monitored throughout the experimental period. pH values of the water during the feeding were measured directly by electronic pH meter (Metler toledo 320 model) by dipping the electrode into each tank. The dissolved oxygen (DO) of the water in experimental tanks was measured using Standardized YSI Do meter (YSI Model 57). The water temperature of the rearing system was measured using a mercury thermometer calibrated from 0°C 110°C on a daily basis. Measurements were carried out by gently immersing the thermometer into the water at vertical position and left for about 2 – 5 minutes. It was quickly moved near the surface of the water and read. Nitrate was measured weekly.

Table 1. Gross composition of experimental diets

Ingredients	Diets (%)				
	D_1(Ctrl)	D_2(DCP)	D_3(CHBM)	D_4(CLBM)	D_5(CABM)
Fishmeal (72%CP)	44.03	36.90	37.30	36.92	36.60
SBM (44%CP)	22.02	18.50	18.50	18.50	18.30
GNC (48%CP)	22.02	18.50	18.50	18.50	18.30
Yellow maize (10%CP)	2.94	2.50	2.50	2.50	2.42
Cod liver oil	5.00	5.00	5.00	5.00	5.00
Vitamin premix	3.00	3.00	3.00	3.00	3.00
Methionine	0.30	0.30	0.30	0.30	0.30
Carboxymethy cellulose	0.70	0.70	0.70	0.70	0.70
Dicalcium phosphate (CaHPO4)	-	1.46	-	-	-
Chicken bone meal (CHBM)	-	-	1.43	-	-
Clarias bone meal (CLBM)	-	-	-	1.46	-
Catttle bone meal (CABM)	-	-	-	-	1.54
Total (%)	100.01	100.00	100.10	100.02	100.02
%Crude protein	40.01	40.00	40.10	40.02	40.02
Available Phosphorus (Calculated) %	0.44	0.80	0.80	0.80	0.80

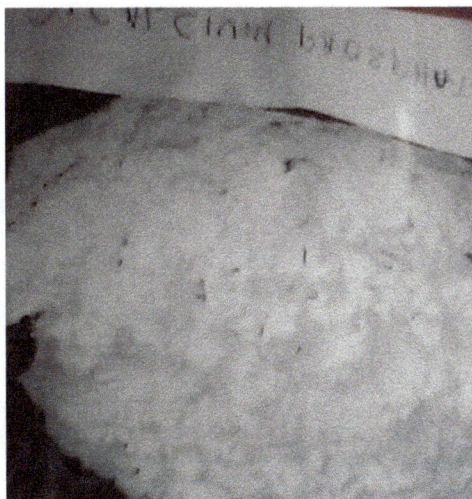

Plate 1. Dicalcium phosphate

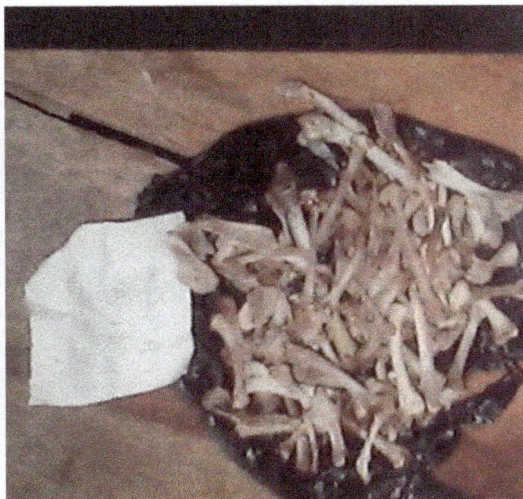

Plate 2. Chicken bone meal

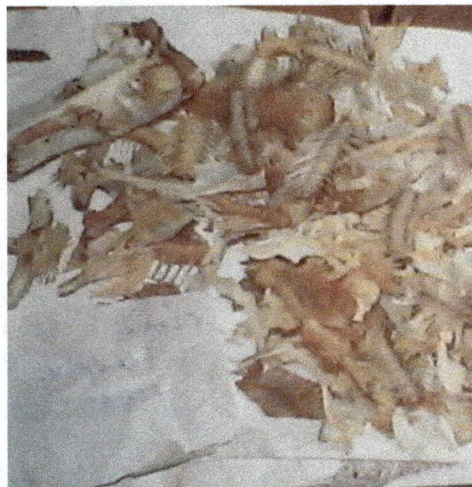

Plate 3. Clarias bone meal

Plate 4. Cattle bone meal

2.4 Growth and Nutrient Utilization Parameters

Growth performance was determined by measuring biweekly mean weight gain (MWG); specific growth rate (SGR) food conversion ratio (FCR), protein efficiency ratio (PER) and net protein utilization (NPU). These growth response parameters were calculated as follow:

MWG (g) = W_i-W_t
Where wt is the final weight of the fish at time (t), w_1 is initial weight of the fish at time 0
SGR (%) = (lnW_t –lnW_i)/T x100
Where W_t is weight of the fish at time t, W_i is weght of the fish at time 0, and T is the culture period in days.

FCR = total dry feed fed (g)/ total wet weight gain (g).
PER = wet weight gain (g)/ amount of protein fed (g)
NPU (%) = 100 x (protein gain/ protein consumed)

2.5 Proximate Composition and Mineral Analysis

Samples of experimental diets and fish (before and after experiment) in all treatments were analyzed for proximate composition and minerals (P and Ca) according to the method of AOAC (2000). Gross energy in kcal/kg of test diet was determined using ballistic oxygen bomb calorimeter (gallen kamp) as described by [12].

2.6 Statistical Analysis

Data obtained from the experiment were subjected to analysis of Variance (ANOVA using SPSS version 15. Differences in means were separated using Duncan Multiple range test.

3. RESULTS

3.1 Proximate Composition of Experimental Diets

The Proximate compositions of experimental diets were presented in Table 2.T The crude protein level of the five diets (D_1-D_5) ranged from 40.07% to 40.19%.The same trend was observed for the nitrogen free extract (NFE) and ether extract. Ash content increased gradually from D_1 (6.10%) and highest in D_4 (6.79). The P content (determined) of the diets ranged between 0.43 and 0.83%. This shows that the diets were formulated in line with the experimental objective. Therefore any changes in the performance of the fish would be attributed to the effect of the supplemented P.

3.2 Proximate Analysis of Experimental Fish after feeding Trial

The results of the proximate analysis are summarized in Table 3. There was no significant difference (P>0.05) in protein content of fish fed D_1 to D_5 when compared with the initial value before feeding trial with the reported values that ranged between (64.25% - 69.05%). Fish fed D_3 (CHBM) had the highest (69.05%) protein content. There was a slight increase in carcass ash with the highest value (15.10%) in fish fed D_5 (CABM). No fibre was detected but there was a

gradual reduction in NFE values in fish carcass before and after the feeding trial as shown in the table.

3.3 Water Quality Parameters

Table 4 shows the mean values of water parameters recorded during the period of the experiment. Water temperature fluctuated within the range of 26.80 – 27.00ºC, while the Dissolved oxygen (DO) ranged from 5.5-6.8 mg/litre. pH values ranged between 6.7-6.9 and nitrate (0.22 – 0.23) mg/L in all the treatments. Though, there was no significant difference (P>0.05) in the values of the parameters analyzed.

3.4 Growth Performance and Nutrient Utilization of Experimental Fish

The growth parameters of the fish fed the various experimental diets were presented in Table 5. Mean weight gain significantly (P>0.05) showed similar pattern to that of SGR with the best values in fish fed D_3 (14.4g, 1.75) respectively. Feed utilization expressed as the feed conversion ratio (FCR) was not significantly (P>0.05) different in fish fed D_2 (1.60) and D_4 (1.53). While the FCR of fish fed D_1 (1.99) and D_5 (1.89) were the poorest. Analyzed values of P and Ca were significantly (P>0.05) highest in fish feed D_3 and D_5 (20.69, 23.79) mg/g respectively.

Plates (5-9) were the samples of fish fed (D_1 –D_5) respectively after the feeding trial. The size of fish in plate 5 (Ctrl) without P supplementation was smallest and biggest in plate 7 (D_3, supplemented with chicken bone meal). Growth of fish in plate 8 and 9 were relatively uniform.

Table 2. Proximate composition experimental diets

Parameters (%)	Diets (%)				
	D_1(Ctrl)	D_2(DCP)	D_3(CHBM)	D_4(CLBM)	D_5(CABM)
Moisture content	12.10	12.13	12.10	12.10	12.12
Crude protein	40.19	40.07	40.12	40.12	40.12
Ether extract	4.21	4.20	4.30	4.30	4.30
Ash	6.10	6.47	6.77	6.79	6.76
Crude fibre	2.25	2.30	2.21	2.23	2.20
NFE	37.40	37.13	36.71	36.69	36.70
phosphorus (determined)	0.43	0.78	0.75	0.82	0.83
Calcium (determined)	1.15	1.19	1.12	1.14	1.15
Ca/P ratio	1-1.1	1-1.2	1-1.2	1-1.1	1-1.2

Table 3. Proximate composition of experimental fish after feeding trial

Parameters(%)	Initial	D_1 (Ctrl)	D_2 (DCP)	D_3 (CHBM)	D_4 (CLBM)	D_5(CABM)
Moisture	11.21±0.15	11.11±0.02	11.20±0.10	11.33±0.06	11.11±0.30	11.34±0.50
Crude protein	64.25±0.07	66.25±0.01	68.03±0.01	69.05±0.07	68.60±0.50	68.01±0.60
Ether Extract	3.58±0.03	3.32±0.02	3.36±0.02	3.33±0.02	4.11±0.50	4.02±0.06
Ash	12.20±0.07	13.60±0.50	13.40±0.07	14.20±0.03	13.60±0.03	15.10±0.04
NFE	5.38±0.04	4.52±0.04	5.21±0.08	3.51±0.50	3.58±0.06	3.81±0.07

Table 4. Mean of water quality parameters recorded during the experimental period

Parameters	D_1 (CTRL)	D_2 (DCP)	D_3 (CHBM)	D_4(CLBM)	D_5 (CABM)
Temperature (°C)	27.00±0.11	26.80±0.21	27.00±0.21	26.80±0.15	26.80±0.20
pH	6.7±0.15	6.9±0.21	6.5±0.20	6.7±0.22	6.7±0.22
DO (mg/L)	6.6±0.20	5.2±0.20	5.1±0.20	4.8 ± 0.22	5.8 ±0.21
Nitrate (mg/L)	0.22±0.20	0.25±0.20	0.24±0.20	0.22±0.20	0.23±0.20

Table 5. Growth performance of *Clarias gariepinus* fed phosphorus supplemented diets

Parameters	D_1(Ctrl)	D_2(DCP)	D_3(CHBM)	D_4(CLBM)	D_5(CABM)
Mean initial body weight(g)	6.00±0.02	6.04±0.01	6.02±0.03	6.01±0.01	6.02±0.03
Mean final body weight(g)	17.67±0.14[b]	18.75±0.12[b]	20.42±0.04[a]	18.75±0.11[b]	18.50±0.05[b]
Mean weight gain(g)	11.67±0.03[b]	12.71±0.20[b]	14.4±0.14[a]	12.74±0.02[b]	12.48±0.05[b]
Average body weight gain daily(g)	0.17±0.01	0.18±0.06	0.21±0.02	0.18±0.03	0.18±0.11
Specific growth rate (SGR)	1.54±0.05[b]	1.62±0.07[b]	1.75±0.05[a]	1.62±0.04[b]	1.62±0.03[b]
Protein efficiency ratio (PER)	0.29±0.02[b]	0.32±0.03[b]	0.36±0.14[a]	0.32±0.06[b]	0.31±0.02[b]
Food conversion ratio (FCR)	1.99±0.02[c]	1.60±0.05[b]	1.25±0.01[a]	1.53±0.05[b]	1.89±0.22[c]
Net protein utilization (NPU)	1.87±0.05	1.89±0.01	1.89±0.12	1.90±0.04	1.88±0.02
Phosphorus (Analyzed) mg/g	12.98±0.20[c]	16.06±0.20[b]	20.69±0.60[a]	16.26±0.21[b]	16.05±0.50[b]
Calcium (Analyzed) mg/g	17.60±0.05[b]	17.37±0.21[b]	23.79±0.21[a]	22.68±0.20[a]	23.80±0.22[a]
Ca/p ratio (analyzed)	1.2:1[b]	1:1[c]	1.1:1[c]	1.5:1[a]	1.2:1[b]

Mean values with similar superscript are not significantly different (P>0.05)

4. DISCUSSION

Evolving recent research works implicate inadequacy of dietary mineral supplementation in fish. It is assumed that fish in their natural habitat should meet the requirements for all the mineral elements. However, the intensive culture of certain fish species in man-made ponds together with dependence on artificial feeding makes it necessary to incorporate adequate quantities of mineral nutrients in the feed. Di-calcium phosphate is an inorganic source of phosphorus, while Chicken bone meal, *Clarias* bone meal and

Cattle bone meal were the organic sources used in this experiment. There is dearth of information on the required level of different bone meals as phosphorus supplement in fish diets. [13] reported a supplementation level of 0.8% P for some selected clariid catfishes using Dicalcium phosphate with 80-90% available phosphorus. Also, [14] reported a supplementation level of 0.7%P using purified diets for *Heterobranchus bidorsalis* fingerlings. In this study, varying bone meal sources as phosphorus supplement in the test diets promoted the growth of *Clarias gariepinus* fingerlings with the best growth

Plate 5. Fish fed with the control diet (0% Phosphorus supplementation)

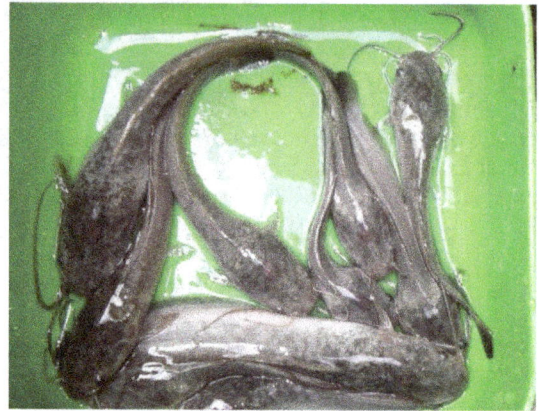

Plate 6. Fish fed with Di-calcium phosphate supplemented diet

Plate 7. Fish fed with Chicken bone meal supplemented diet

Plate 8. Fish fed with Clarias bone meal supplemented diet

Plate 9. Fish fed with cattle bone meal supplemented diet

performance in fish fed Chicken bone meal-supplemented diet.

Phosphorus is an essential component of organic compounds involved in almost every aspect of metabolism. For optimum performance, P needs to be present in the diet at required levels. The dietary requirement for phosphorus ranges between 0.45 - 1% [5]. However, in some cases feeds do not contain sufficient amounts of P and supplementation is necessary to optimize fish performance. Fresh water fish require dietary sources of phosphorus to meet their relatively high metabolic requirement since the water is deficient in phosphorus [15].

5. CONCLUSION

There is need for phosphorus supplementation in the diets of young indigenous catfish such as *Clarias gariepinus*. The study revealed poor growth performance at no P supplementation, while body phosphorus continued to increase above that required for maximum growth in fish fed cattle bone meal- supplemented diet. The study had shown that phosphorus is a growth promoter when supplied from the right source and at optimal concentration in fish feed which validated some of the early work done on warm water fresh fishes. Under practical condition, organic phosphorus sources should be included in the diets to prevent poor growth and symptoms of skeletal deformity often experienced in artificially raised fish. In addition, organic sources of phosphorus in fish feed are cheaper and easy to source and this will help reduce cost of feed formulation.

COMPETING INTERESTS

Authors have declared that no competing interests exist.

REFERENCES

1. NRC. National research council. Nutrient requirements of warm water fishes and shell fishes. National Academy Press, Washington DC, USA. 1993;114.

2. Kyeong-Jun Lee, Madison S, Powell, Frederick T, Barrows, Scott Smiley, Peter Bechtel, Ronald W. Hardy. Evaluation of supplemental fish bone meal made from Alaska seafood processing by products and dicalcium phosphate in plant protein based diets for rainbow trout (*Oncorhynchus mykiss*). Aquaculture. 2010;302:248-255.

3. Davis DA, Gatlin DM. Dietary requirements of fish and marine crustaceans. Rev. Fish Sci. 1996;4:75-99.

4. Rania S, Mabroke Azab M, Taboun Ashraf Suloma, Ehab R, El-Haroun. Evaluation of

meat and bone meal and mono-sodium phosphate as supplemental dietary phosphorus sources for broodstocks nile tilapia (*Oreochromis niloticus*) under the condition of hapa-in- pond system. Turkish Journal of Fisheries and Aquatic Sciences 2013;13:11-18.

5. Lall SP. The minerals. In: Fish nutrition. 3rd edition. Halver JE, Hardy RW. eds. academic press, New York. 2002;264–274.

6. Omitoyin BO. Introduction to fish farm in Nigeria. University press. Ibadan, Nigeria; 2007.

7. Lall SP. Comparative mineral nutrition of fish sources and requirements. In: Nutritional biotechnology in the feed and food industries, Lyons TP, Jacques KA, JM Hower (Eds.). Nottingham University Press, Nottingham, UK. 2007;387-391.

8. Ye CX, Liu YJ, Tian LX, Mai KS, Du ZY, Yang HJ, Niu J. Effect of dietary calcium and phosphorus on growth, feed efficiency, mineral content and body composition of juvenile grouper. Epinephelus Coioides Aquacult. 2006;255:263-267.

9. Asgard T, Shearer KD. Dietary phosphorus requirement of juvenile Atlantic salmon, Salmon Salar. C. Aquaculture Nutr. 1997;3:17-23.

10. AOAC. Association of Analytical chemists. Official methods of analysis 16th edition. 2000;1141.

11. Faturoti EO, Balogun AM, Ugwu LLC. Nutrient utilization and growth responses of *Clarias* (Clarias lazera) fed different dietary protein levels. Nigerians Journal of Applied Fisheries and Hydrobiology. 1986;1:41-45.

12. AOAC. Association of official analytical chemists. Official methods of analysis. Washington DC. USA. 15th edition. 1990;200-210.

13. Adebayo IA. Growth and blood chemistry of selected clariid catfishes fed phosphorus supplemented diets. Journal of Fisheries and Aquatic Science. 2011;6(5):545-554.

14. Adebayo IA. Optimal phosphorus requirement of *Heterobranchus bidorsalis* using purified diet. International journal of Agriculture and forestry. 2012;2(5):195-198.

15. Albrektsen S, Hope B, Aksnes A. Phosphorus (P) deficiency due to low P availability in fishmeal produced from blue whiting (*Micromesistius poutassou*) in feed for under-yearling Atlantic salmon (*Salmo salar*) smolt. Aquaculture. 2009;296:3118-328.

Growth Performance, Carcass Characteristics and Economic Efficiency of Using Graded Levels of Moringa Leaf Meal in Feeding Weaner Pigs

A. D. Oduro-Owusu[1], J. K. Kagya-Agyemang[1], S. Y. Annor[1] and F. R. K. Bonsu[1*]

[1]Department of Animal Science Education, University of Education, Winneba, Box 40, Mampong-Ashanti, Ghana.

Authors' contributions

This work was carried out in collaboration between all authors. Author ADOO, JKKA and FRKB designed the study, wrote the protocol, managed the analyses of the study and wrote the manuscript. Author SYA reviewed the experimental design and performed the statistical analysis. All authors read and approved the final manuscript.

Editor(s):
(1) Hugo Daniel Solana, Department of Biological Science, National University of Central Buenos Aires, Argentina.
Reviewers:
(1) Andell Edwards, Animal Science Department, University of Trinidad and Tobago, Trinidad and Tobago.
(2) Emad Shaker, Biochemistry Dept., Minia Univ., Egypt.

ABSTRACT

A study was conducted to evaluate the feeding value of *Moringa oleifera* leaf meal (MOLM) as part of feed ingredient on the growth performance, carcass characteristics, and economic efficiency of weaner pigs. A total of forty-five (45) weaner pigs of mixed sexes of age 7-8 weeks old were allocated to five dietary treatments and nine replicates in a randomized complete block design. The treatments were: diet 1 designated as 0% MOLM had no moringa in the diet and was the control, diets 2, 3, 4, and 5 designated as 1% MOLM, 2.5% MOLM, 3.5% MOLM and 5% MOLM contained moringa leaf meal at 1%, 2.5%, 3.5% and 5% respectively. Data collected were subjected to analysis of variance with the aid of SAS (2008). The results obtained showed that feed intake and final body weight were not significantly ($p>0.05$) influenced by MOLM. The growth rate of pigs on 5% MOLM (0.54 kg/pig) was better ($p<0.05$) than those on the control and 2.5% MOLM diets and this reflected in the best feed conversion efficiency (0.3) for the pigs on 5% MOLM. Carcass parameters including slaughter weight, organ weight, carcass length, loin eye muscle area, ham and primal cuts of pork were not significantly ($p>0.05$) influenced by MOLM. Back fat thickness

---

Corresponding author: E-mail: fritzobonsu@yahoo.com;

reduced (p<0.05) from 2.2 cm in the control to 1.7 cm as moringa inclusion increased to 5%. There were no differences in crude protein levels of the meat (20.2% to 24.6%), moisture content (69.1% to 71.3%), and the pH of the meat (5.3 to 6.0). The feed cost decreased as the level of MOLM inclusion in the dietary treatments increased from 0% MOLM to 5% MOLM. It was therefore concluded that MOLM could be used as a feed ingredient in the diet of pigs to reduce production cost. MOLM had no detrimental effect on the meat of pigs, and has the potential to reduce fat level in pork to produce leaner carcass.

Keywords: Moringa oleifera; pigs; growth performance; carcass characteristics; leaf meal.

1. INTRODUCTION

Pig production provides the means by which rapid transformation of animal protein consumption can be achieved in Ghana. Although pigs are frequently maligned by some social and religious groups in Ghana, they have several good attributes including high prolificacy, high fecundity, short generation interval, early maturity, high feed conversion efficiency and a modest requirement with respect to housing and equipment [1]. Despite these advantages of pig production, high cost of feed increases the price of pork beyond the reach of the average Ghanaian. The urgent need to improving production efficiency, through lower production costs and supply of a product that meets consumers' expectations are key elements required for a profitable and viable pig production enterprise [2]. Lowering the feed cost demands the use of non-conventional feedstuffs that are readily available with good nutrient composition for use in pigs' diets.

Moringa oleifera Lam. (moringa) which is a promising non-leguminous multipurpose tree with high crude protein and lower tannins content [3,4,5] offers a good alternative source of protein to animals. Moringa is a fast-growing tree with fast regrowth after pruning [6] and has the capacity to produce high quantities of fresh biomass per square meter even at high planting densities. Moringa is a source of highly digestible protein (methionine and cystine), calcium, iron, ascorbic acid, and carotenoids [4,6,7]. In spite of these good attributes of moringa, there is scanty information regarding its use in pig diets.

This study was therefore undertaken to determine the optimum level at which Moringa oliefera leaf meal (MOLM) could be incorporated in the diets of weaner pigs to improve their growth performance, carcass characteristics and economic efficiency.

2. MATERIALS AND METHODS

2.1 Experimental Location

The experiment was carried out at the Piggery Section of the Department of Animal Science Education, University of Education, Winneba (UEW), Mampong-Ashanti. The experiment lasted for six months.

2.2 Experimental Pigs and Design

Forty-five (45) large white weaner pigs of age 7-8 weeks old and mixed sexes (30 females and 15 males) obtained from the Piggery Section of the Animal Science Farm, University of Education, Winneba, Mampong-Ashanti were used for the experiment. The animals were balanced by weight and allocated to five dietary treatments and nine replications in randomized complete block design. Each animal constituted a replicate. The five dietary treatments were: Diet 1, which was designated as 0% MOLM, was the control diet and contained soya bean meal and fish meal as the main protein source with no moringa leaf meal. Diet 2 designated as 1% MOLM, Diet 3 as 2.5% MOLM, Diet 4 as 3.5% MOLM and Diet 5 as 5% MOLM contained moringa leaf meal at the rate of 1%, 2.5%, 3.5% and 5%, respectively. The dietary treatments were iso-nitrogenous and iso-carloric. The proportion of the individual feed ingredients in the dietary treatments is presented in Table 1.

2.3 Housing and Feeding

Each animal was housed singly in a pen (194 cm x 160 cm) with concrete floor that had a feeder and a drinker. The experimental diets were offered ad libitum in separate concrete feeders in the morning (07.00 h). The diets were offered daily and the leftover feed weighed daily before feeding. Water was also provided ad libitum.

Table 1. Composition of experimental diets and calculated analysis

Feed ingredients	% Composition of ingredients per treatment (As Is)				
	0% MOLM	1% MOLM	2.5% MOLM	3.5% MOLM	5% MOLM
Maize (grain)	50	50	50	50	50
Anchovey fish meal	2.0	2.0	2.0	2.0	2.0
Tuna fish meal	5.5	5.0	5.5	4.0	5.0
Soybean meal	7.7	7.5	6.5	7.5	6.0
Wheat bran	34.3	34.5	33.5	33.0	32.0
Premix	0.5	0	0	0	0
Moringa leaf meal	0	1.0	2.5	3.5	5.0
Total	100	100	100	100	100
Calculated analyses					
Crude protein (%)	17.26	17.27	17.31	17.29	17.33
Crude fibre (%)	4.5	4.7	4.8	5.1	5.2
Ether extract (%)	3.88	3.87	3.91	3.91	3.98
DE (MJKg^{-1})	13.1	13.1	13.2	13.0	12.9

MOLM- Moringa oleifera leaf meal, DE- Digestible Energy

*Vitamin premix provided the following per kilogram of diet: Fe 100 mg, Mn 110 mg, Cu 20 mg, Zn 100 mg, Se 0.2 mg, Co 0.6 mg, Senoquin 0.6 mg, retinal 2000 mg, cholecalciferol 25 mg, α- tocopherol 25 mg, menadione 1.33 mg, cobalamin 0.03 mg, thiamin 0.83 mg, riboflavin 2 mg, folic acid 0.33 mg, biotin 0.03 mg, pantothenic acid 3.75 mg, macin 23.3 mg, pyridoxine 1.33 mg

2.4 Parameters Measured

Parameters measured included growth performance, carcass characteristics and economics of production using moringa based diets.

2.5 Data Analysis

Data obtained were subjected to analysis of variance with [8]. Differences among means were separated using least significant difference (LSD) at 5% significant level.

3. RESULTS AND DISCUSSION

3.1 Proximate Composition of Moringa Leaves

The MOLM contained a considerable quantity of crude protein, crude fibre, ash, and ether extract (Table 2). The MOLM used in this study had a crude protein content of 26.70% (Table 2) which was lower than 40.00%, 30.30% and 27.40% reported by [9-11] respectively but was higher than values of 16.00%, 22.42% and 23.27% obtained by [12-14] respectively. The crude fibre value of 14.63% obtained was lower than the 19.25% and 19.10% reported by [11] and [6] respectively. The ether extract value of 5.00% was lower than the 6.50% reported by [10] but similar to the 5.25% reported by [6]. The ash content of 9.00% was also lower than the

12.00% reported by [15] but the dry matter content of 88.75% was higher than the value of 76.53% reported by [11]. These variations in the nutrient contents of moringa in this study and others reported may be due to differences in agro-climatic conditions or to different ages of trees [6], age of cutting or harvesting, edaphic factors, agronomic practices as well as methods of processing and analysis of MOLM [16].

Table 2. Proximate composition (%) of moringa leaf meal

Parameters	Value (%)
Crude protein	26.70
Crude fibre	14.63
Ether extracts	5.00
Ash	9.00
Dry matter	88.75
Moisture	11.25
Nitrogen free extracts	33.42

3.2 Growth Performance of the Pigs

The growth performance of weaner pigs fed on the dietary treatments for the experimental period is shown in Table 3.

Initial body weights of the pigs were similar (p>0.05) at the start of the experiment. Daily feed intake was not influenced (p>0.05) by inclusion of MOLM in the diets. A major factor influencing feed intake in pig is the energy density of the diet when physiological and environmental factors are held constant [17,18]. The general similarities of feed intake observed in this study among the treatments indicated the iso-carloric nature of the dietary treatments (Table 1) as pigs ate to satisfy their energy requirements [17]. The growth rate of pigs on 5% MOLM was better (p<0.05) than those fed on the control diet and 2.5% MOLM diets but not significantly different from pigs on 1% MOLM and 3.5% MOLM. The final body weight was not influenced by dietary treatments. The feed conversion efficiency (FCE) was better (p<0.05) for pigs fed on 5% MOLM (0.30 kg wt/kg feed) as compared to the control and 2.5% MOLM (0.28kg wt/kg feed) but not different (p>0.05) from 1% MOLM and 3.5% MOLM. The variations in feed conversion efficiency are attributed to differences in the feed utilization by the animals, although diets were iso-nitrogenous. The observations made on growth rate and FCE further suggest that MOLM can be used in pig diets to obtain good growth performance without vitamin premix inclusion in the diet. One of the main constraints for the use of tropical foliages could be the high content of the fibre fractions and bulkiness [19]. These constraints did not adversely affect the growth performance of the pigs at 5% inclusion level and indicated potential higher levels of MOLM inclusion in the feed.

3.3 Carcass Characteristics

The mean carcass weight, organ weight and dressing percentage of pigs are presented in Table 4. Significant differences (P<0.05) were observed in the carcass dressed weights among dietary treatments and is consistent with observation made by [20]. Whereas carcass dress weight of 0% MOLM, 1% MOLM, 2.5% MOLM and 5% MOLM were similar (p>0.05) those pigs fed the diet that contained 3.5%

MOLM had higher (p<0.05) dressed weight as compared with 0% MOLM and 2.5% MOLM. However, Dressed percentage was not influenced (P>0.05) by MOLM. Muscle score was significantly (p<0.05) better for 5% MOLM diet probably due to comparative nutrient availability and utilization by the animals. Carcass characters such as loin muscle area, carcass length, ham, and feet were not influenced significantly (P>0.05) by dietary treatments. This indicates that the use of MOLM in pig diets is not likely to adversely affect these carcass traits. Back fat thickness was lower (p<0.05) for pigs fed 5% MOLM (1.7 cm) as compared with the control (2.2 cm) and 1% MOLM (2.1 cm) but not 2.5% MOLM and 3.5% MOLM (2 cm) and has the advantage of producing lean carcass. The trend indicates an inverse relationship between the level of MOLM inclusion and back fat thickness. The reduction in back fat is attributed to the MOLM which possesses a potent hypocholesterolemic agent [21] that probably reduced the fat composition of the body. This further suggests that when pigs are fed on MOLM diet, it is likely to reduce the overall fat content of the resulting meat.

The weight of lungs, heart, spleen, intestines and kidneys was not significantly (p>0.05) influenced by MOLM in the diets (Table 4). The weight of the liver of pigs fed the control diet was higher (p<0.05) as compared with the 2.5% MOLM but not the other treatments. The comparative lower liver weight could be due to the hypocholesterolemic property of MOLM which might have reduced the fat built up in the liver. These results indicate that MOLM is not likely to cause any detrimental effect to the carcass traits or organs.

Table 3. Growth performance of weaner pigs fed moringa leaf meal in kg

Parameters	0% MOLM	1% MOLM	2.5% MOLM	3.5% MOLM	5% MOLM	LSD	SE
Initial body weight (kg/pig)	14.2	14.4	14.3	14.1	14.2	0.34	0.16
Daily feed intake (kg/pig)	1.79	1.77	1.75	1.77	1.77	0.06	0.03
Total feed intake (kg/pig)	164.9	162.8	161.5	162.3	162.8	10.78	5.02
Final body weight (kg/pig)	60.7	60.2	60.2	60.1	60.2	0.62	0.29
Total weight gain (kg/pig)	46.5	45.8	45.9	46.0	46.0	0.69	0.33
Daily growth rate (kg/day)	0.51[b]	0.52[ab]	0.50[b]	0.52[ab]	0.54[a]	0.03	0.01
Feed conversion efficiency	0.28[b]	0.29[ab]	0.28[b]	0.29[ab]	0.30[a]	0.02	0.01

Means bearing different superscript in the same row are significantly different (p<0.05)
MOLM = Moringa leaf meal; LSD=Least significant difference; SE =Standard error

Table 4. Effect of different levels of moringa leaf meal (MOLM) on the carcass components of pigs

Parameters	0% MOLM	1% MOLM	2.5% MOLM	3.5% MOLM	5% MOLM	LSD	SE
Slaughter weight (kg)	60.7	60.2	60.2	60.1	60.2	0.62	0.29
Dress weight (kg)	51.2[b]	51.9[ab]	50.9[b]	55.1[a]	54.2[ab]	3.62	1.10
Dressing percentage (%)	84.3	86.2	84.5	91.6	90.0	0.36	2.0
Last rib fat thickness (cm)	2.0	1.4	2.2	2.0	1.5	1.11	0.34
Muscle score	2.0[b]	2.0[b]	2.0[b]	2.0[b]	3.0[a]	0.0	0.0
Loin muscle area (cm^2)	17.0	17.5	18.0	16.6	16.3	1.31	0.40
Carcass length (cm)	61.7	63.0	59.0	63.0	67.0	8.22	2.52
Ham (kg)	7.9	6.9	7.6	7.7	8.2	1.82	0.55
Shoulder (kg)	7.0[b]	7.3[b]	7.3[b]	8.1[ab]	8.7[a]	1.22	0.37
Back fat thickness (cm)	2.2[a]	2.1[a]	2.0[ab]	2.0[ab]	1.7[b]	0.44	0.13
Feet (kg)	0.7	1.2	1.2	1.0	1.2	0.42	0.13
Full gastro intestinal tract (%LBW)	7.4	6.7	6.7	6.9	6.5	1.61	0.49
Empty gastro intestinal tract (%LBW)	3.6	3.4	3.7	3.7	3.7	1.18	0.36
Liver (%LBW)	1.9[a]	1.7[ab]	1.3[b]	1.8[ab]	1.7[ab]	0.59	0.18
Lungs (%LBW)	0.8	0.9	1.2	0.8	0.7	0.55	0.16
Heart (%LBW)	0.4	0.4	0.4	0.4	0.3	0.18	0.05
Spleen (%LBW)	0.2	0.2	0.1	0.2	0.1	0.22	0.07
Kidney (%LBW)	0.3	0.3	0.3	0.4	0.3	0.21	0.06

Means bearing the same superscript in the same row are not significantly different (p>0.05)
LSD= Least significant difference; SE= Standard error
%LBW= Percentages of live body weight

3.4 Meat Characteristics

The protein, ether extract, moisture and pH of pork from the various treatments are indicated in Table 5.

The mean values for protein, ether extract, moisture and hydrogen potential were not influenced (p>0.05) by dietary treatments. The similar chemical composition of the meat observed could possibly be attributed to the fact that the protein content of the diets was iso-nitrogenous. Protein content of a diet is directly related to the moisture level of the carcass, which also affects the ether extract level [22]. This indicates that diets that contained MOLM were as good as the control. The ultimate pH is of particular importance to the meat industry because it directly influences the self-life, colour and eating quality of meat [23,24]. The desirable pH for meat ranges from 5.5 to 5.8 and it is associated with light-coloured and tender meat [25,26]. The pH of meat has a high influence on water holding capacity (WHC), which is closely related to product yield and pork quality. The pH of the meat obtained in this study 5.2, 5.3, 5.3 and 6.2 for 3.5% MOLM, 1% MOLM, 5% MOLM and 2.5% MOLM respectively were not within the

recommended range but the control was within the accepted range. Low pH is usually associated with Pale, Soft and Exudative (PSE) pork and is not desirable. On the other hand, high meat pH (above 6.0 to 6.2) often causes dark, firm and dry (DFD) pork. The pH values obtained in this study could be attributed to pre-slaughter stress on the pigs.

3.5 Economic Efficiency of Using MOLM

Cost efficiency of using MOLM in pig diets is presented in Table 6. The per kg feed cost reduced as the MOLM inclusion increased. Total feed cost also decreased as the level of MOLM in the diets increased from 0% MOLM to 5% MOLM (Table 6). The diets containing 1% MOLM, 2.5% MOLM, 3.5% MOLM and 5% MOLM contained more of the less expensive MOLM and relatively less amounts of other protein ingredients (soya bean meal and fish meal) which are more expensive. This is consistent with observation made by [27]. At the time of the experiment, the production cost of MOLM was GH¢0.20 per kg while the prevailing market prices of soya bean meal and fish meal were GH ¢4.00 and GH¢ 2.00 per kg, respectively.

Table 5. Effect of different levels of MOLM on the carcass components of weaner pigs (dry matter)

Parameters	0% MOLM	1% MOLM	2.5% MOLM	3.5% MOLM	5% MOLM	LSD	SE
Protein (%)	20.2	21.9	23.3	24.6	22.8	11.36	4.20
Ether extract (%)	23.1	15.8	18.9	14.3	21.7	11.85	4.12
Moisture (%)	71.3	70.3	70.2	69.0	69.1	5.96	2.10
pH	5.5	5.3	6.2	5.2	5.3	4.29	4.80

Means bearing the same superscript in the same row are not significantly different (P>0.05)
LSD= Least significant difference, pH= Hydrogen potential, SE = Standard error of means

Table 6. Economic efficiency of using MOLM

Parameters	0% MOLM	1% MOLM	2.5% MOLM	3.5% MOLM	5% MOLM
Per kg feed cost (GH¢)	0.938	0.915	0.9015	0.899	0.883
Total feed intake (kg)	164.9	162.8	161.5	162.2	162.0
Total feed cost (GH¢)	154.6	148.9	145.6	145.8	143.0
Weight gain (kg)	46.5	45.7	45.8	46.0	46.0
Feed cost: weight gain (GH¢/kg)	3.32:1	3.26:1	3.18:1	3.17:1	3.10:1

The cost to gain ratio ranged from 3.10:1 to 3.32:1. Pigs fed diets containing 5% MOLM had the best feed cost to gain ratio of 3.10:1 whiles the poorest was the diet that had no moringa (0% MOLM- 3.32:1). Thus, diet containing 5% MOLM was more economical than any of the other dietary treatments and could be attributed to higher weight gain (46.0 kg) with respect to total feed cost (GH¢143.0). This indicates that the inclusion of MOLM in pig diets renders the production more economical.

4. CONCLUSION

The results of this study show that moringa leaf meal (MOLM) has good nutrient composition particularly protein which could be used as a feed ingredient among others for feeding weaner pigs. The growth performance and general carcass parameters were not adversely affected by the inclusion of MOLM in pigs' diet. It is apparent that MOLM has a potential to reduce back fat thickness of pigs. It was more economical to produce pigs especially at 5% inclusion level. Further research is needed to establish the effect of MOLM on sensory analysis and haematological indices of the pigs.

COMPETING INTERESTS

Authors have declared that no competing interests exist.

REFERENCES

1. Okai DB, Abora PKB, Davis T, Martin A. Nutrient composition, availability, Current and potential uses of "Dusa": A cereal by-product obtained from "koko" (porridge) production. J. of Sci. & Techno. 2005;25:33-38.
2. Mullan B, D'Souza DN. The role of organic minerals in modern pig production. In: Redefining Mineral Nutrition. (J. A. Tayloy Pickard and L. A Tucker, eds). Nottingham University press, UK. 2005;89-106.
3. Morton JF. The horseradish tree, *Moringa pterygosperma* (Moringaceae) - A boon to arid lands? Econ. Bot. 1991;45:318-333.
4. Makkar HPS, Becker K. Nutritional value and anti-nutritional components of whole and ethanol extracted *Moringa oleifera* leaves. Anim. Feed Sci. & Techno. 1996;63:211-228.
5. Gohl B. Tropical feeds. FAO. Rome; 1998. Available:http://www.fao.org/ag/aga/agap/frg/afris/default.htm
6. Foidl N, Makkar HPS, Becker K. The potential of *Moringa oleifera* for agricultural and industrial uses. In: Fugile, L.J. (ed). The miracle tree: The multiple attributes of Moringa. CTA Publication. Wageningen, The Netherlands. 2001;45-76.
7. Fahey W, Zakmann AT, Talalay P. The chemical diversity and distribution of glucosinolates and isothiocyanates among plants phytochemistry. 2001;56(1):5-51

8. Statistical Analysis System (SAS). User's Guide, SAS/STAT® 9.2, Cary, NC:SAS Institute Inc; 2008.

9. Sanchez-Machado DI, Nunez-Gastelum JA, Reyes-Moreno C, Ramirez-Wong B, Lopenz-Cervantes J. Nutritional Quality of edible Parts of *Moringa oleifera*. Food Anal. Methods. 2009;3:175-180.

10. Moyo B, Patrick J, Masika AH, Voster M. Nutritional characterization of Moringa (*Moringa oleifera* Lam.) leaves. African J. of Biotech. 2011;10:12925-12933.

11. Oduro I, Ellis WO, Owusu D. Nutritional potential of two leafy vegetables: *Moringa oleifera* and *Ipomoea batatas* leaves. Sci. Res. Essays. 2008;3(2):57-60.

12. Gidamis AB, Panga JT, Sarwatt SV, Chove BE, Shayo NB. Nutrients and anti-nutrient contents in raw and cooked leaves and mature pods of *Moringa oleifera*, Lam. Ecol. Food Nutr. 2003;42:399-411.

13. Sarwatt SV, Milang'ha MS, Lekule FP, Madalla N. *Moringa oleifera* and cotton seed cake as supplements for smallholder dairy cows fed napier grass. Livestock Resear for Rural Develo. 2004;16:38.

14. Nouala FS, Akinbamijo OO, Adewum A, Hoffman E, Muetzel S, Becker K. The influence of *Moringa oleifera* leaves as substitute to conventional concentrate on the *In vitro* gas production and digestibility of groundnut hay. Livest Res. Rural Dev. 2006;18:Article #121.

15. Gupta K, Barat GK, Wagle DS, Chawla HKL. Nutrient contents and antinutritional factors in conventional and non-conventional leafy vegetables. Food Chem. 1989;31: 105-116.

16. Fuglie L. Producing food without pesticides: local solution to crop pest control in West Africa. CTA, Wageningen, The Netherlands; 1999.

17. NRC (National Research Council). Nutrient Requirements of Swine. 10th ed. National Academy Press, Washington, DC; 1998

18. Noblet J, Van Milgen J. Energy value of pig feeds: Effect of pig body weight and energy evaluation system. J. Anim. Sci. 2004; 82 (13):229-238.

19. García M. Use of four grading of foliage and root dehydrated sweet potato (*Ipomoea batatas* L.) in pigs, from the stage of growth and its effect on the productive behavior. Thesis. Faculty of Agronomy. Central University of Venezuela. Maracay; 1998.

20. Gadzirayi CT, Masamha B, Mupangwa J F, Washaya S. Performance of broiler chickens fed on mature *Moringa oleifera* leaf meal as a protein supplement to soyabean meal. Int. J. of Poult. Sci. 2012;11:5-10.

21. Ghasi S, Nwobodo E, Ofili JO. Hypocholesterolemic effects of crude extract of leaf of *Moringa oleifera* Lam in high-fat diet fed Wister rats. J. of Ethnopharm. 2000;69:21-25.

22. Noblet J, Henry Y. Energy evaluation systems for pigs diets: A review. Livest. Prod. Sci. 1993;36:121-141

23. Fernandez X, Tornberg E. A review of the causes of variation in muscle glycogen content and ultimate pH in pigs. J. Muscle Foods. 1991;2:209–235.

24. Przybylski W, Vernin P, Monin G. Relationship between glycolytic potential and ultimate pH in bovine, porcine and ovine muscles. J. Muscle Foods. 1994;5:245–255.

25. Gardner GE, Kenny L, Milton JTB, Pethick DW. Glycogen metabolism and ultimate pH in Merino, first cross and second cross wether lambs as affected by stress before slaughter". Austr. J. of Agric. Res. 1999;50:175-18.

26. Bidner BS, Ellis M, Brewer MS, Campion D, Wilson ER, McKeith FK. Effect of ultimate pH on the quality characteristics of pork. J of Muscl. Foods. 2004;5:139-154

27. Zanu HK, Asiedu P, Tampuori M, Abada M, Asante I. Possibilities of using moringa (*Moringa oleifera*) leaf meal as a partial substitute for fishmeal in broiler chickens diet. Online J. Anim. Res. 2012;2(1):70-75.

6

The Importance of Legumes in the Ethiopian Farming System and Overall Economy: An Overview

Mulugeta Atnaf[1,2*], Kassahun Tesfaye[2] and Kifle Dagne[2]

[1]Ethiopian Institute of Agricultural Research, Pawe research center, P.O.Box 25, Pawe, Ethiopia.
[2]Department of Microbial, Cellular and Molecular Biology, Collage of Natural Sciences, Addis Ababa University, P.O.Box 1176, Addis Ababa, Ethiopia.

Authors' contributions

This review work was carried out in collaboration between all authors. Author MA prepared the paper design and setup and all drafts of the manuscript, and managed the statistical analyses of the review. Author KT and KD framed and edited the manuscript. All authors read and approved the final manuscript.

Editor(s):
(1) Jamal Alrusheidat, Extension Education Department, National Centre for Agricultural Research and Extension (NCARE), Amman, Jordan.
(2) Chrysanthi Charatsari, Department of Agricultural Economics, School of Agriculture, Aristotle University of Thessaloniki, Greece.
(3) Daniele De Wrachien, State University of Milan, Italy.
Reviewers:
(1) Michael N. I. Lokuruka, Food Science and Nutrition, Karatina University, Kenya.
(2) Anonymous, USA.
(3) Anonymous, Ethiopia.
(4) Anonymous, Brazil.
(5) Anonymous, India.

ABSTRACT

Crops, livestock and trees are major components of farming systems in Ethiopia. Crop production is dominant in Ethiopian agriculture as well as in the farming system. Legumes are among the various crops produced in all regions of the country in different volumes after cereals. More than twelve legume species are grown in the country. Pulses production by volume has been increased by 71.92% for the duration of nearly 20 years and with a growth rate of 3.78% per annum. Area coverage by pulse crops for the same period grown by 53% with a growth rate of 3% per year. Total pulses grain yield, which is volume of production per unit area, showed good increment from 8.79 quintals per hectare in the cropping year 1994/1995 to 14.76 quintals per hectare in

Corresponding author: E-mail: atnafmulugeta@gmail.com;

2012/2013 cropping season. However, it is much lower compared to the potential demonstrated in research managed fields. Legumes have multiple uses. Grain legumes provide food and feed and facilitate soil nutrient management. Herbaceous and tree legumes can restore soil fertility and prevent land degradation while improving crop and livestock productivity sustainably. The pulse industry in the country has developed significantly with little intervention, and great potential exists to increase the production and impact of pulses through proactive and targeted support. The role that Ethiopia now plays in the international pulse market can be attributed to significant growth rates in pulse production over the last nearly 20 years. However, bunch of constraints and considerable gaps lean the legumes along the value-chain from production to marketing and utilization. The country needs to target the constraints and gaps to optimize the importance of legumes in the farming system and economy of the country.

Keywords: Ethiopia; farming system; legumes/pulses; sustainability.

1. INTRODUCTION

Ethiopia has diverse agro-ecology that permits different farming systems. Crops, livestock and trees are major components of the farming systems in the country [1]. Existence of diverse farming systems, socio-economic, cultures and agro-ecologies, endowed Ethiopia with a diverse biological wealth of plants, animals, and microbial species, especially crop diversity. Crop plants such as coffee (*Coffeea arabica),* safflower (*Carthamus tinctorius*), tef (*Eragrostis tef),* noug (*Guizotia abyssinica),* anchote (*Coccinia abyssinica)* and enset *(Ensete ventricosum)* originated in Ethiopia. High genetic diversity is found in major food crops (wheat, barley, sorghum and peas); industrial crops (linseed, castor and cotton); cash crop (coffee); food crops of regional and local importance (tef, noug, Ethiopian mustard, enset, finger millet, cowpea, lentil) and in a number of forage species of world importance such as clovers, medics, and oats [2].

Though the diverse agro-ecological setting permits diverse farming and livelihood systems, crop production is by far the largest component of agriculture in particular and the country's economy in general [3]. Out of the total arable land of 50.5 million hectares, close to 16.4 million hectares are suitable for producing annual and perennial crops. Of this estimated land area, about 8 million hectares (nearly 50%) are used annually for rain-fed small holder crops [4 -11].

Food crops can be categorized as cereals, pulses, oilseeds among others for simplicity of description and comparison purposes [11]. According to CSA (2012/2013), Out of the total grain crop area, 78.17% (9.6 million hectares) is covered under cereals. Teff, maize, wheat, sorghum and barley took up the lion share of cereals area coverage and volume of grain production. Pulses are among the various crops produced in all the regions in different volumes across the country after cereals. Twelve and more legume species are grown in the country. Of these, faba bean (Vicia *faba* L.), field pea (*Pisum sativum* L.), chickpea (*Cicer arietinum* L.), lentil (*Lens cultinaris* Medik.), grass pea (*Lathyrus sativus* L.), fenugreek (*Trigonellafoenum-graecum* L.) and lupin (*Lupinus albus L.*) are categorized as high to mid land pulses. On the other hand, haricot bean (*Phaseolus vulgaris* L.), soya bean (*Glycine max* L.), cowpea (*Vigna unguiculata* L.), pigeon pea (*Cajanus cajan* L.) and mung bean are predominantly grown in the warmer and low land parts of the country [11].

Pulses production by volume has been increased by 71.92% for the duration of nearly 20 years and with a growth rate of 3.78% per annum (Fig. 1). Area coverage by pulses crops for the same period has been increased by 53% with a growth rate of 3% per year (Fig. 2). Total pulses grain yield, which is volume of production per unit area, showed significant increment from 8.79 quintals per hectare in the cropping year 1994/1995 to 14.76 quintals per hectare in 2012/2013 cropping season (Fig. 3). However, it is much lower compared to the potential demonstrated in research managed fields (Fig. 7). Area coverage and productivity are the two most important factors of production. From these data, one can safely deduce that the increment of volume of pulses production was attributed to both of the above mentioned factors. Currently, the national average productivity of soybean, chick pea, faba bean, and grass pea is greater than the productivity of the pulses as a group.

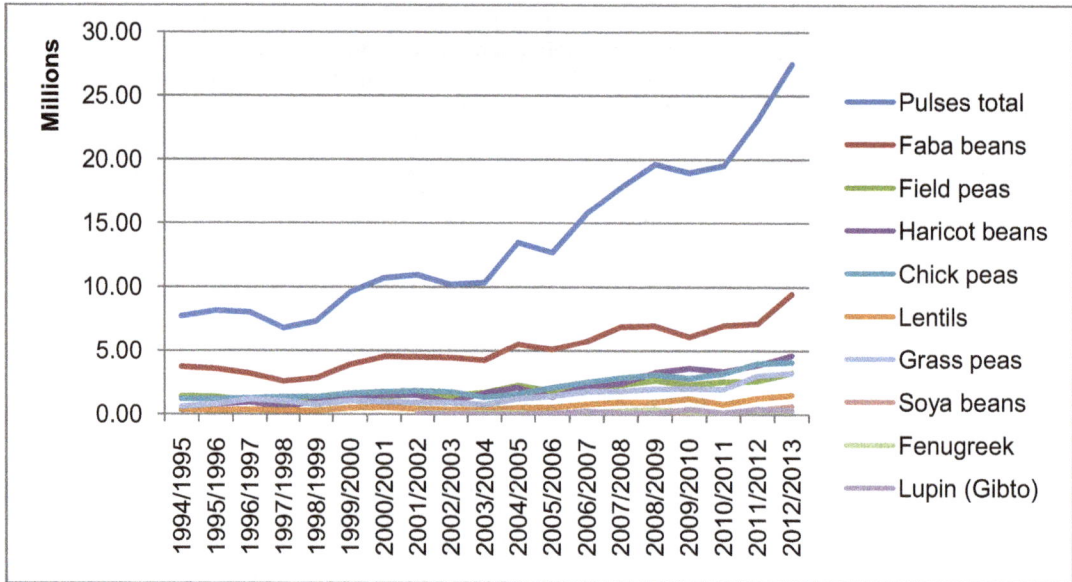

Fig. 1. Nineteen cropping years (1994/95-2012/2013) volume of production (in million quintals) trend of different pulses crops in Ethiopia (Source: [4-22])

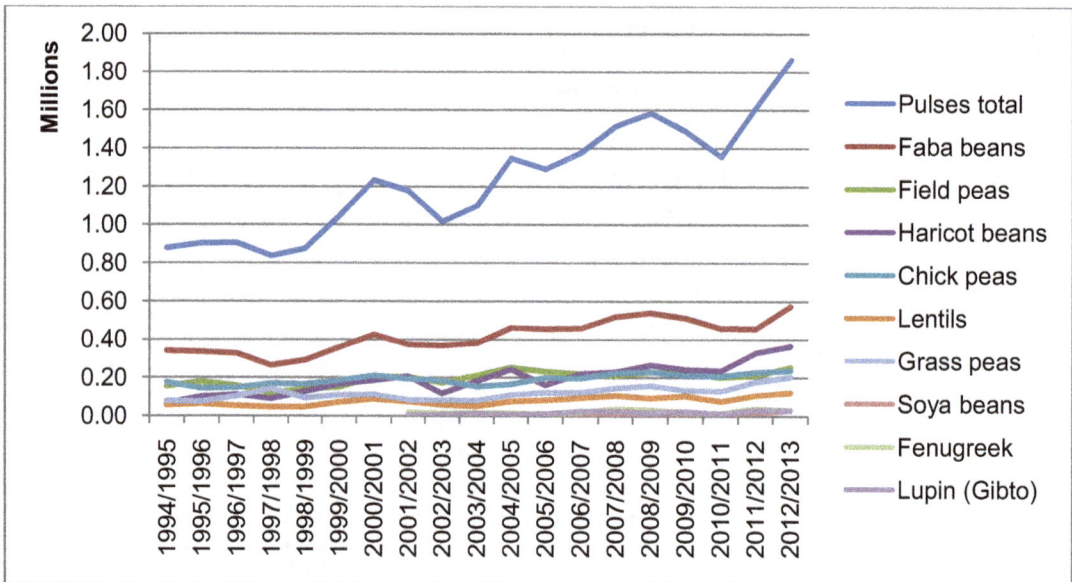

Fig. 2. Nineteen cropping years (1994/95-2012/2013) area coverage (in million hectares) trend of different pulses crops in Ethiopia (Source: [4-22])

2. IMPORTANCE OF LEGUMES

Legumes are known to perform multiple functions. Grain legumes provide food and feed, facilitate soil nutrient management and contributes to climate change mitigation [23]. Herbaceous and tree legumes can restore soil fertility and prevent land degradation while improving crop and livestock productivity on a more sustainable basis. Thus cultivation of such dual-purpose legumes, which enhance agricultural productivity while conserving the natural resource base, may be instrumental for achieving income and food security, and for reversing land degradation [24].

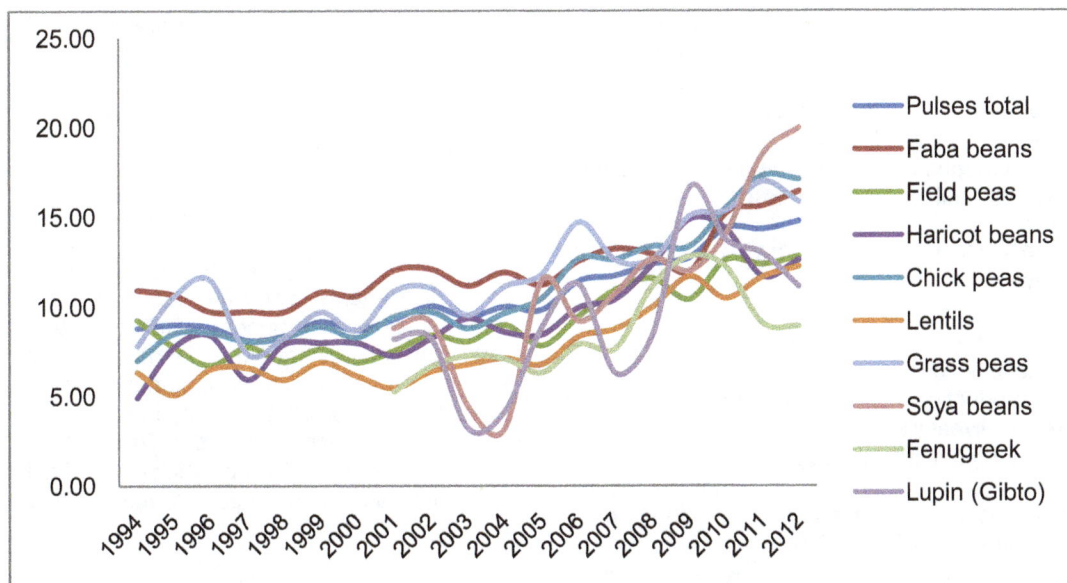

Fig. 3. Nineteen cropping years (1994/95 - 2012/2013) productivity (measured in quintals per hectare) trend of different pulses crops in Ethiopia (Source: [4-22])

Ethiopian farmers' produce different legume crops mainly for food and feed, to fetch cash, and more importantly to restore the fertility of the crop land. Participation of farmers on cultivation of pulses in the country has been increased nearly by double from 4.5 to 8.5 million farmers for the last nearly 20 years (Fig. 4). Legumes contribute to smallholder income; as a higher-value crop than cereals and to diet, as a cost effective source of protein that accounts for approximately 15 percent of protein intake. Moreover, pulses offer natural soil maintenance benefits through nitrogen-fixing, which improves yields of cereals through crop rotation, and can also help smallholder farmers reduce cost of inorganic fertilizers. It also contribute significantly to Ethiopia's balance of payments [25].

2.1 Farming System Importance of Legumes

One very important offer of legumes is that it improves soil and environmental health. Biological nitrogen fixation (BNF) is the distinguishing feature of a legume in a cropping system. Most legume species are able to form a symbiosis with alpha- or beta proteo-bacteria, collectively called rhizobia, that use solar energy captured by the plant to break the bond in inert atmospheric di-nitrogen and form reactive N species, initially ammonium (NH_4^+). As a result of this symbiosis, the legume crop requires little or no input of N fertilizer and makes little demand

on soil N reserves [23,26,27]. For example, lupin can potentially fix and accumulate a total of 150 to 400 kg/ha per year nitrogen [28,29,30] and faba bean up to 200 kg/hectare per year. The benefits of using legumes such as soybean and peanut as break crops in sugar cane monoculture on the wet tropical coast of Queensland are well documented with yield increases of the following cane crop in the order of 20–30% [31,32]. In addition to reducing the levels of cane pests and diseases [33], soybean rotations also provide economic benefits from the harvested grain and economic and environmental benefits from the N-rich residues reducing subsequent fertilizer N inputs [32]. Similarly, other legumes play a vital role in controlling major cereal root diseases, particularly cereal eelworm or cereal cyst nematodes, *Heterodera avenae* in the Mediterranean region. The combination of high soil N and reduced nematodes population is cumulative and can result in a big increase in subsequent cereal yield [34].

Since the manufacture of synthetic fertilizer consumes fossil fuel, thereby releasing CO_2, and the transport and spreading of organic and synthetic N fertilizers consumes further fuel, the use of legumes in cropping systems has immediate environmental benefits arising from reduced fossil fuel use. Nitrates from fertilizers and soil N reserves may also leach through the soil column into groundwater, and the denitrification of nitrates from synthetic or organic sources is the primary source of nitrous oxide (N2O), a powerful greenhouse gas, from agricultural soils [35]. Hence maintaining the reactive N within the plant, as happens in a symbiotic legume in the growing season, avoids some potential for environmental damage.

2.2 Importance of Legumes to Livestock Feed

Legumes have been shown to improve both the quantity and quality of fodder, and thus sustain feed production during the dry season and increase livestock productivity. Experiments in Ethiopian highlands showed that forage legumes did not reduce the barley grain and straw yield, but significantly increased the total fodder yield – barley straw plus forage [36]; similar results were found for maize [37]. Average fodder yields of 14.2 and 3.4 tons per hectare of maize-vetch and barley-clover, respectively, were reported compared to 9.3 and 2.3 tons per hectare of sole maize and barley, respectively [38]. The average crude protein content of crop residues is about 3.8% of dry matter, whereas legumes crude protein content on average varies between 14-24% of dry matter [39]. In Ethiopia, Crossbred cows given an oats-vetch diet produced on average 1.40 kg/day more milk than those given hay diet (5.54 vs. 4.14 kg milk/day) [40]. Legumes mixed with crop residues also increase

other livestock production parameters like weaning rate, manure and lactation yield [24].

2.3 Importance of Legumes for Human Nutrition and Health

Legumes are plant-based alternatives to animal products with low health impact. They are characterized by a high content of protein, fiber and micronutrients and low fat. The protein content of legumes is generally between 20-30% which corresponds to levels found in fish and meat. Low lipid intake is important for good health. The total oil content in legume seeds was the lowest, compared with that in seeds and grains of other crops. Except chick peas (5 g/100 g), the total oil contents of other legumes are less than 1.6 g/100 g.

The levels of saturated fatty acids are also very low in all types of legumes. Moreover, there are high levels of unsaturated fatty acids (>70% of total oil). The content of unsaturated fatty acids in chickpeas is up to 4.3 g/100 g. Low saturated fatty acids are identified as an important standard for evaluation of the quality of food [41]. Legume seeds contain low levels of total oil and saturated fatty acids, as well as high content of unsaturated fatty acids, therefore increase intake of legumes can be beneficial to human health [42,43]. For instance, lupin grain is uniquely high in protein

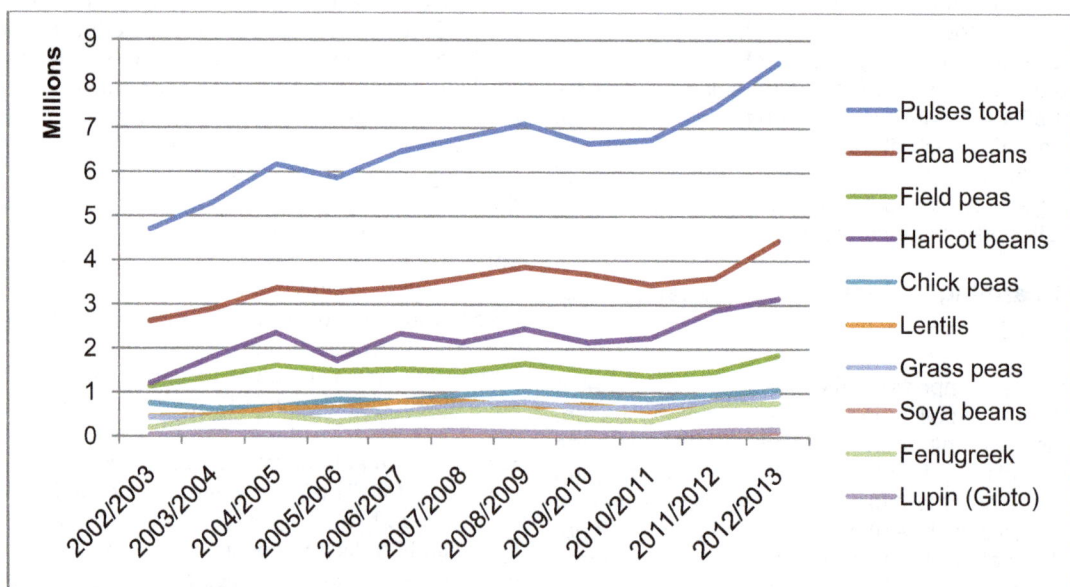

Fig. 4. Number of participant farmers who cultivated different pulses crops during the last 11 years (2002/2003 - 2012/2013) in Ethiopia (Source: [4-11 & 19-22])

(30 to 40 per cent) and dietary fiber (30 per cent) and low in fat (6 per cent). It contains minimal starch and, therefore, has a very low glycemic index which has significant implications for modern societies with an increasing incidence of obesity and associated risk of diabetes and cardiovascular disease [44]. In nutritional terms, lupin seed is an attractive alternative to soybean for human consumption. Food laboratory studies have shown that the protein and fiber components have excellent functional properties [42]. Lupin ingredients have been included in a range of highly palatable breads and other baked goods, meat products and beverages [45]. Studies have also indicated the substantial health attributes of lupin such as diets supplemented with lupin grain may play an important role in treating type 2 diabetes, particularly in overweight and obese people, beneficially influence satiety (appetite suppression) and energy balance, improve blood lipids, lower hypertension and improve bowel health [45,46,47,48].

2.4 Contribution of Legumes to Small Holder Livelihood

Pulses contribute to smallholder livelihoods in multiple ways. Firstly, it can play a significant role in improving smallholders' food and nutrition security, as an affordable source of protein (pulses make up approximately 15 percent of the average Ethiopian diet) and other essential nutrients like potassium, iron and zinc. As a protein source, pulses are more affordable than meat, fish, and dairy products for smallholders, and for more than 43 percent of Ethiopians who practice Orthodox Christianity, pulses become the single largest source of protein during the fasting period. Secondly, pulses can have an income benefit for smallholders, both in terms of diversification and because they yield a higher gross margin than cereals. The income benefits of diversifying from cereals to pulses for one smallholder farmer are exemplified in a case study considering the net profit of wheat, barley and teff, compared to faba beans and chickpeas in Ethiopia. The results demonstrate that pulses are generally more profitable than cereals, giving smallholders an economic incentive to increase pulse production. Faba beans provide the highest net return among the crops considered, while chickpeas provide higher returns than barley and teff, but comparable returns to wheat [25].

2.5 Wider Economic Perspectives

Legume comprise different important commodities such as haricot bean, chick pea, faba bean etc both in domestic and export markets, in the Ethiopian trade of balance. The export value of legumes in the Ethiopian economy has been increased by 89.92% over the last 18 years with 5% annual growth rate (Fig. 5). The actual export value by legumes in 1995 was 20.34 million USD and 201.86 million USD in the year 2012. This export value increment have a positive impact on the trade balance, and contribute to the country's foreign exchange reserves. The national export value as whole has been increased during the same period from 568 million USD in 1995 to 2.99 billion USD in 2012. The export share of legumes to the national export value during the period mentioned has been increased thought fluctuated from 3.59% in 1995 as minimum to 8.2% in 2002 as maximum and a little bit lower (6.73%) in 2012 (Fig. 6).

3. POTENTIALS AND OPPORTUNITIES

The pulse industry has developed significantly with little intervention, and great potential exists to increase the production and impact of pulses through proactive and targeted support. Rough calculations suggest that Ethiopia could expand its foreign market presence by at least doubling its current exports through increased production levels [25]. Smallholder income could also be increased by at least 40-70 percent per hectare of pulses planted through greater pulse productivity (with better inputs and sound agronomic practices) [25]. There is an opportunity to stabilize and increase supply by improving production up to the full potential which would meet domestic demands, helping to ensure food security.

Ethiopia is now one of the top twelve producers of total legumes in the world, the second-largest producer of faba beans after China, and the fifth or sixth largest producer of chickpea. The role that Ethiopia now plays in the international pulse market can be attributed to significant growth rates in pulse production over the last nearly 20 years. For instance, in 1994/5 the country produced only 742 thousand tons of pulses, compared to 1.95 million tons in 2010/11 and 2.75 million tons in 2012/2013 cropping year [11].

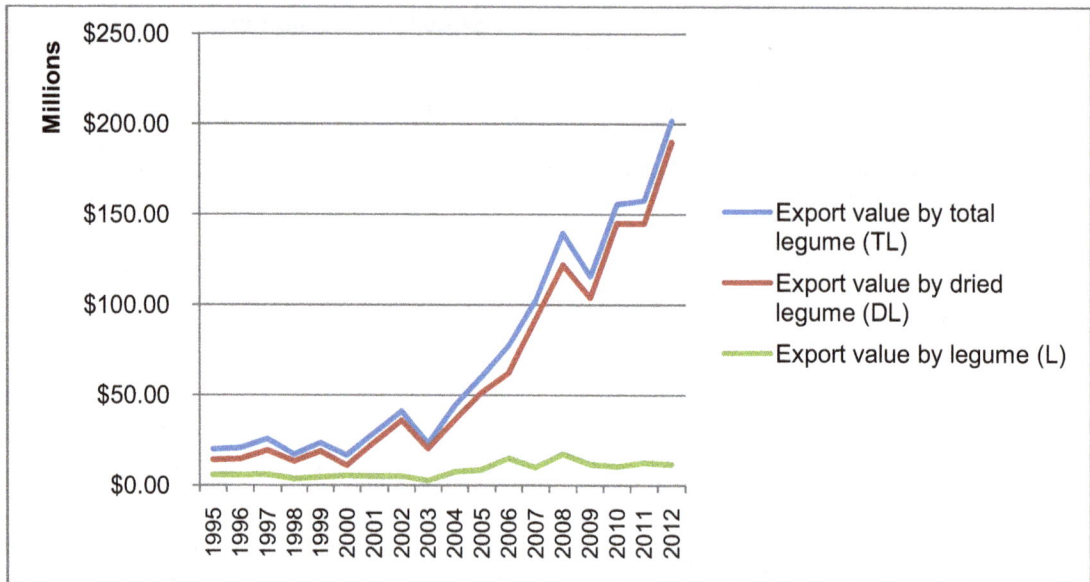

Fig. 5. Export value trend of legumes in the Ethiopian export market over the last 18 years (Source: [49])

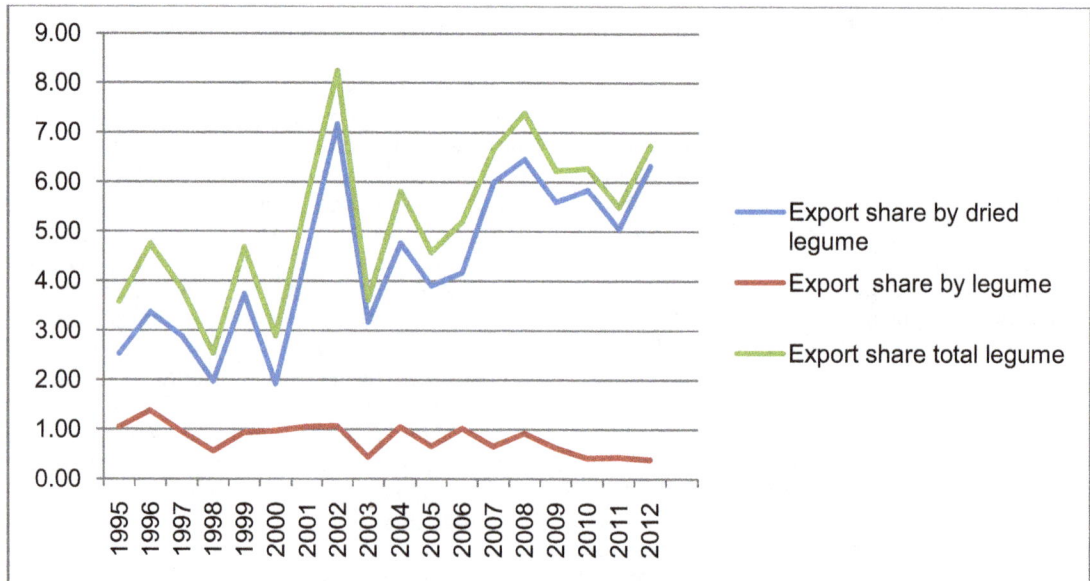

Fig. 6. Percent export share trend of legumes in the Ethiopian export market over the last 18 years (Source: [49])

Through productivity and market improvements, the critical role of legumes in smallholder livelihood and food security can be expanded. The current productivity of pulses falls significantly below the demonstrated potential (Table 1). For example, current average haricot bean yields are 12 quintals per hectare, but research demonstrated potential in Ethiopia is 34 quintals per hectare if accompanied by the appropriate inputs. This gain in productivity would not only increase smallholder income by 40 to 70 percent per hectare, but would also ensure greater food security through meeting domestic pulse demand. In addition, Ethiopia could expand its foreign market presence through increased production levels, which will lead to at least doubling of its current annual exports.

4. GAPS AND CHALLENGES

A number of constraints and considerable gaps prevail the legumes along the value-chain from production to utilization. Generally, legumes have got less attention in terms of crop management, and input utilization by different development actors' especially small scale farmers compared to cereals. This gap is mainly attributed to the perception of the farmers. The low grain productivity per unit area of legumes compared to cereals is a real challenge that legume scientists should always look into. Assessments in Ethiopia show, productivity is below potential due to: low input usage; limited availability of seed and limited familiarity with the variety of existing legumes, and limited usage of modern agronomic practices and poor extension services (Table 1).

The national agricultural research system managed the development and release of 169 improved varieties of different food legumes in the country during the periods 1973 to 2012 (Fig. 8). On average, the system delivered four varieties per annum. However, there are neglected potential pulses that the research system should look in to and develop improved and standard technologies. Lupin, grass pea, mung bean and fenugreek are good examples.

Moreover, the research system should continually deliver improved standard technologies for every potential legumes significantly produced in the country.

Numerous constrains have been identified in assessments made in the country regarding marketing and export of legumes. Weak linkage between the producers and the export markets have been identified as a major constrain. The weakness is due to the large number of ineffective intermediaries operating in the value chain who have failed to acquire scale and operate in limited geographic areas. The fragmentation of intermediaries between the producer and consumer markets creates a lack of transparency in markets. While there has been substantial growth in recent years, the current export market is underdeveloped. The less developed and fragmented exporters operating at smaller scale in the market results in inconsistent export flows and thus, inconsistent demand for exports [25]. Major causes of limited export development identified were: (i) inadequate market intelligence (ii) inability to leverage scale efficiencies due to smaller size and (iii) non-conducive business environment due to missing credit and insurance; and (iv) inconsistent policy interventions [25].

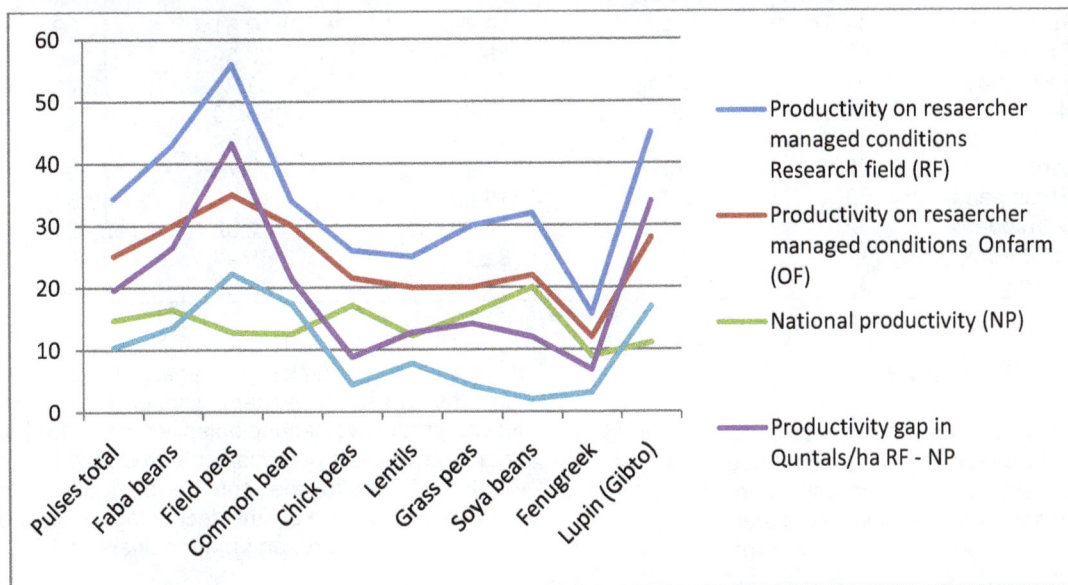

Fig. 7. Productivity gap between research demonstrated potential and national average of food legumes

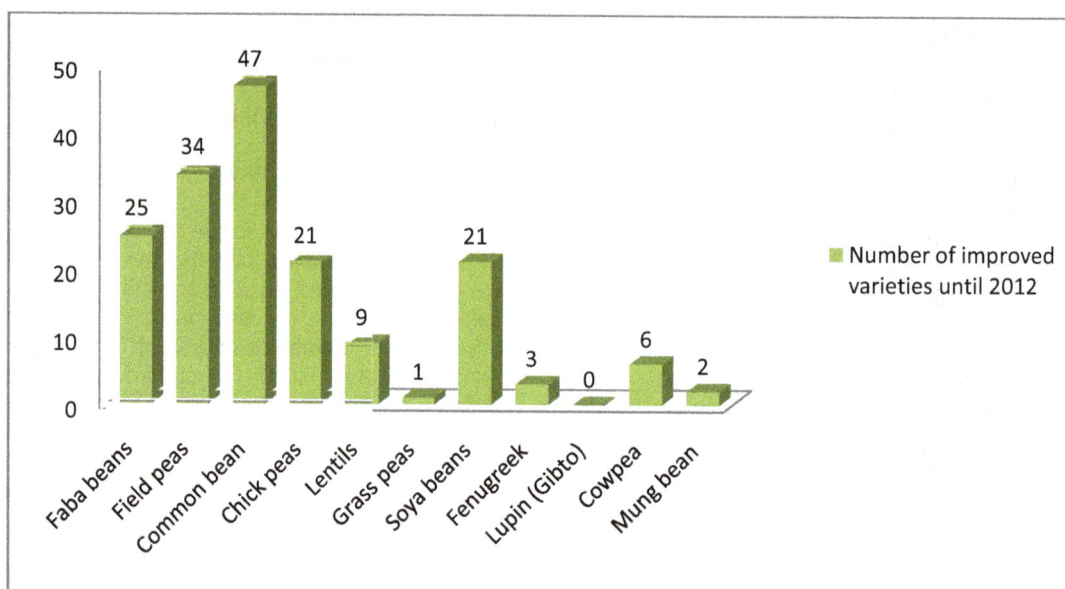

Fig. 8. Number of improved varieties of food legumes released by national agricultural research system of Ethiopia during the periods 1973 - 2012 (Source: [50,51])

Table 1. Productivity gap between research demonstrated potential and national average in the year 2012 for the different food legumes

Legume species	Productivity on research managed conditions		National productivity (NP), 2012	Productivity gap in Quintals/ha	
	Research field (RF)	On-farm (OF)		RF - NP	OF - NP
Pulses total	34.37	21.15	14.76	19.61	10.39
Faba beans	43	30	16.44	26.56	13.56
Field peas	56	35	12.79	32.21	22.21
Haricot beans	34	30	12.62	20.38	12.38
Chick peas	26	21.5	17.11	10.89	5.89
Lentils	25	20	12.25	12.75	7.75
Grass peas	35	30	15.85	14.15	4.15
Soya beans	32	22	19.98	12.02	2.02
Fenugreek	15.7	12	8.91	6.79	3.09
Lupin (Gibto)	45	28	11.12	33.88	16.88

Source [50,51]

5. CONCLUSION

Legumes could offer multiple uses. It could be used as human food, animal feed, improves soil and environmental health, and as a human nutrition and medicine. Moreover, it is good and cheap source of dietary protein and fetch reasonable cash for the Ethiopian poor farmers. The country has huge potential to produce different legumes and to benefit from their multiple uses. Nevertheless, the country needs to give sufficient attention to the sub-sector in terms of changing the perception of the farmers to produce up to the demonstrated potential by using all the production packages. In line with this, the research system should proactively deliver productive technologies for the different legume crops that potentially be produced in the country. To make the country competitive in potential world market, the technologies should be up to the standard, and market oriented.

ACKNOWLEDGMENTS

Authors would like to thank synonymous reviewers and Mr. Demeke Mewa for their significant contribution towards the improvement of the manuscript.

COMPETING INTERESTS

Authors have declared that no competing interests exist.

REFERENCES

1. African Development Bank Group (ADBG), Ethiopian Review of Bank Group Assistance to Agricultural and Rural Development Sector; 2008.
2. Institute of Biodiversity Conservation. Second country report on the state of PGRFA To FAO: Ethiopia; 2007.
3. Temesgen T, Reshid MH. Economic impact of climate change on crop production in Ethiopia; Evidence from Cross Section Measures. Journal of African Economies. 2005;18(4):529-554.
4. Ethiopian Central Statistical Agency (ECSA). Report on area and production of crops (Private peasant holdings, Meher season). Addis Ababa, Ethiopia; 2006.
5. Ethiopian Central Statistical Agency (ECSA). Report on area and production of crops (Private peasant holdings, Meher season). Addis Ababa, Ethiopia; 2007.
6. Ethiopian Central Statistical Agency (ECSA). Report on area and production of crops (Private peasant holdings, Meher season). Addis Ababa, Ethiopia; 2008.
7. Ethiopian Central Statistical Agency (ECSA). Report on area and production of crops (Private peasant holdings, Meher season). Addis Ababa, Ethiopia; 2009.
8. Ethiopian Central Statistical Agency (ECSA). Report on area and production of crops (Private peasant holdings, Meher season). Addis Ababa, Ethiopia; 2010.
9. Ethiopian Central Statistical Agency (ECSA). Report on area and production of crops (Private peasant holdings, Meher season). Addis Ababa, Ethiopia; 2011.
10. Ethiopian Central Statistical Agency (ECSA). Report on area and production of crops (Private peasant holdings, Meher season). Addis Ababa, Ethiopia; 2012.
11. Ethiopian Central Statistical Agency (ECSA). Report on area and production of crops (Private peasant holdings, Meher season). Addis Ababa, Ethiopia; 2013.
12. Ethiopian Central Statistical Agency (ECSA). Report on area and production of crops (Private peasant holdings, Meher season). Addis Ababa, Ethiopia; 1995.
13. Ethiopian Central Statistical Agency (ECSA). Report on area and production of crops (Private peasant holdings, Meher season). Addis Ababa, Ethiopia; 1996.
14. Ethiopian Central Statistical Agency (ECSA). Report on area and production of crops (Private peasant holdings, Meher season). Addis Ababa, Ethiopia; 1997.
15. Ethiopian Central Statistical Agency (ECSA). Report on area and production of crops (Private peasant holdings, Meher season). Addis Ababa, Ethiopia; 1998.
16. Ethiopian Central Statistical Agency (ECSA). Report on area and production of crops (Private peasant holdings, Meher season). Addis Ababa, Ethiopia; 1999.
17. Ethiopian Central Statistical Agency (ECSA). Report on area and production of crops (Private peasant holdings, Meher season). Addis Ababa, Ethiopia; 2000.
18. Ethiopian Central Statistical Agency (ECSA). Report on area and production of crops (Private peasant holdings, Meher season). Addis Ababa, Ethiopia; 2001.
19. Ethiopian Central Statistical Agency (ECSA). Report on area and production of crops (Private peasant holdings, Meher season). Addis Ababa, Ethiopia; 2002.
20. Ethiopian Central Statistical Agency (ECSA). Report on area and production of crops (Private peasant holdings, Meher season). Addis Ababa, Ethiopia; 2003.
21. Ethiopian Central Statistical Agency (ECSA). Report on area and production of crops (Private peasant holdings, Meher season). Addis Ababa, Ethiopia; 2004.
22. Ethiopian Central Statistical Agency (ECSA). Report on area and production of crops (Private peasant holdings, Meher season). Addis Ababa, Ethiopia; 2005.
23. Baddeley JA, Jones S, Topp CFE, Watson CA, Helming J, Stoddard FL. Biological nitrogen fixation (BNF) in Europe. Legume Futures Report 1.5; 2013.
    Available: www.legumefutures.de
24. Kassie M. Economic and Environmental Benefits of Forage Legume-Cereal Intercropping in the Mixed Farming System: A Case Study in West Gojam, Ethiopia. Addis Ababa, Ethiopia: EDRI; 2011.
25. Rashid S, Yirga C, Behute B, Lemma S. Pulses value chain in Ethiopia: Constraints and opportunities for Enhancing exports: IFPRI; 2010.
26. Gemechu K, Endashaw B, Fassil A, Muhammad I, Tolessa D, Kifle D, Emana G. Evaluation of Ethiopian chickpea (*Cicer arietinum* L.) germplasm accessions for

symbio-agronomic performance. Renewable Agriculture and Food Systems; 2012. DOI:10.1017/S1742170512000221.

27. Mcvicar R, Panchuk K, Pearse P. Soil Improvements with Legumes. Saskatchewan Agriculture; 2005. Accessed on October 2014. Available: http://www.agriculture.gov.sk.ca/

28. Takunov IP Yagovenko LL. Yellow lupin (*Lupinus luteus* L.) as a green manure crop preceding winter rye (*Secale cereale* L.). In: E van Santen, Wink M, Weissmann S Roemer P (Editors). Lupin, an ancient crop for the new millennium. Proceedings of the 9th International Lupin Conference, Klink/Muritz, 20-24 June 1999. International Lupin Association, Canterbury, New Zealand. 1999;434-437.

29. Reeves DW, Touchton JT, Kingery RC. The use of lupin in sustainable agriculture systems in the southern coastal plain. Abstracts of technical papers, No. 17, Southern Branch ASA, Little Rock, USA; 1990.

30. Jansen PCM. *Lupinus albus* L. [Internet] Protabase. Brink, M. & Belay, G. (Editeurs).PROTA (Plant Resources of Tropical Africa, Wageningen, Pays Bas; 2006. Accessed On May 2014. Available:http://Database.Prota.Org/Recherche.Htm

31. Garside AL, Bell MJ The value of legume breaks to the sugarcane cropping system – cumulative yields for the next cycle, potential cash returns from the legume, and duration of the break effect. Proceedings of the Australian Society of Sugar Cane Technologists. 2007;29:299-308.

32. Garside AL, Bell MJ. Fallow legumes in the Australian sugar industry: Review of recent research findings and implications for the sugarcane cropping system. Proceedings of the Australian Society of Sugar Cane Technologists. 2001;23:230-235.

33. Pankhurst CE, Stirling GR, Magarey RC, Blair BL, Holt JA, Bell MJ, Garside AL. Quantification of the effects of rotation breaks on soil biological properties and their impact on yield decline in sugarcane. Soil Biology & Biochemistry. 2005;37:1121-1130.

34. Ates S, Feindel D, El Moneim A, Ryan J. Annual forage legumes in dryland agricultural systems of the West Asia and North Africa Regions: research achievements and future perspective. Annual forage legumes in dryland agricultural systems. 2013;69:17-31.

35. Philippot L, Hallin S. Towards food, feed and energy crops mitigating climate change. Trends in Plant Science. 2011;16:476-480.

36. Zewdu T. Grain and straw yield of barley as influenced by under sowing time of annual forage legumes and fertilization. Tropical Science. 2004;44:85-88.

37. Zewdu T. Effect of defoliation and intercropping with forage legumes on maize yield and forage production. Tropical Science. 2003;43:204-207.

38. Zewdu T, Assefa G, Mengistu A. The role of forages and pastures crops for increased and sustainable livestock production. Research procedure, past achievements and future directions in North-Western Ethiopia. Research Report. Ethiopia, Adet Agricultural Research Center; 2000.

39. Mpairwe DR, Sabiiti NE, Ummuna NN, Tegegne A, Osuji P. Integration of legumes with cereal crops I. Effect of supplementation with graded levels of lablab hay on voluntary food intake, digestibility, milk yield and milk composition of crossbred cows fed maize-lablab Stover or oats-vetch hay ad-libitum. Livestock Production Science. 2003;79:193-212.

40. Khalili H, Osuji PO, Ummunna NN, Crosse S. The effects of forage type (maize lablab or oat-vetch) and level of supplementation (wheat-middling) on food intake, diet apparent digestibility, purine excretion and milk production of Crossbred Cows (BosTaurus × Bosindicus). Animal Production; 1992;58:183-189.

41. Shi Feng. Legume. Food and Wine Sciences Group, Lincoln University; 2008. Accessed on September 2014. Available:http://foodscience.wikispaces.com/Legume#Grain%20legumes

42. Lqari H, Pedroche J, Girion-Calle J, Vioque J. Production of *Lupinus angustifolius* protein hydrolysates with improved functional properties. Fats and Oils. 2005;56(2):135-140.

43. Ryan E, Galvin K, O'Connor TP, Maguire AR,. O'Brien NM. Phytosterol, squalene, tocopherol content and fatty acid profile of selected seeds, grains and legumes. Plant Foods Hum. Nut. 2007;62(3):85-91.

44. Hall RS, et al. Australian sweet lupin flour addition reduced the glycaemic index of a

white bread breakfast without affecting palatability in healthy human volunteers. Asia Pacific J. Clinical Nutrition. 2005;14:91-97.

45.   Hall RS, Johnson SK, Baxter AL, Ball MJ. Lupin kernel fibre-enriched foods beneficially modify serum lipids in men. European Journal of Clinical Nutrition. 2005;59:325-33.

46.   Archer BJ et al. Effect of fat replacement by inulin or lupin-kernel fibre on sausage patty acceptability, post-meal perceptions of satiety and food intake in men. British Journal Nutrition. 2005;91:591-599.

47.   Capraro J, Clemente A, Rubio LA, Magni C, Scarafoni A, Duranti M. Assessment of the lupin seed glucose-lowering protein intestinal absorption by using in vitro and ex vitro models. Food chemistry. 2010;10-073.

48.   Johnson SK, Chua V, Hall RS, Baxter AL. Lupin kernel fibre foods improve bowel function and beneficially modify some putative faecal risk factors for colon cancer in men. British Journal of Nutrition. 2006;95(2):372-8.

49.   The Observatory of Economic Complexity (OEC): Products Exported by Ethiopia. Accessed on July 31, 2104. Available:http://atlas.media.mit.edu/explore/tree_map/hs/export/eth/all/show/2012/

50.   Ministry of Agriculture, Animal and Plant Health Regulatory Directorate. Crop variety register. Addis Ababa, Ethiopia. 2012;15.

51.   Ministry of Agriculture and Rural Development, Crop Development Department. Crop variety register. Addis Ababa, Ethiopia. 2005;15.

7

Impact of Financial Sector Reforms on Agricultural Growth in Nigeria: A Vector Autoregressive (Var) Approach

Aniekan Jim Akpaeti[1*]

[1]Department of Agricultural Economics and Extension, Akwa Ibom State University, Ikot Akpaden, Mkpat Enin, P.M.B.1167, Uyo, Akwa Ibom State, Nigeria.

Author's contribution

Author AJA was responsible for sourcing, analysis, typing of data and proofreading of manuscript.

Editor(s):
(1) Anthony N. Rezitis, Dept. of Business Administration of Food and Agricultural Enterprises, University of Patras, Greece.
(2) Daniele De Wrachien, State University of Milan, Italy.
Reviewers:
(1) Ionel-Mugurel Jitea, Economic Sci., University of Agricultural Sciences and Veterinary Medicine Cluj-Napoca, Romania.
(2) Anonymous, Nigeria.
(3) Anonymous, USA.
(4) Anonymous, Ghana.
(5) Alex A. A. Bruce, Business Administration, Gombe State University, Nigeria.

ABSTRACT

The financial sector has witnessed several reforms over the decade. The associated impact of this is also felt in the agricultural sector. The study was carried out to assess the impact of financial sector reforms on agricultural growth in Nigeria from 1970-2009. Secondary data were collected from Central Bank of Nigeria, National Bureau of Statistics and National Population Commission and analyzed using vector error correction model (VECM) approach. The result revealed that financial sector reforms in the baseline and sensitivity model significantly impact on agricultural growth both in the long and short-run. However, the impact of financial sector reforms shock in the sensitivity model on agricultural output growth was lower by 0.60 percent when compared with 78.85 percent in the baseline model. This implied that financial sector reforms could play a significant role in the growth of the agricultural sector by increasing its production level and independently generate positive investments in the sector than in the sensitivity result. It is therefore recommended that government should adopt strong macroeconomic policies targeted to kick-start meaningful growth in the agricultural and financial sectors as well as provide the enabling environment for farming as a business through concessionary interest rates, tax free and import duty concessions. These

*Corresponding author: E-mail: anigreat04@yahoo.com;

financial and fiscal incentives when provided would encourage further output growth in the agricultural sector of the country.

Keywords: Agricultural growth; financial sector reforms; impact; Nigeria.

1. INTRODUCTION

The issue of financial sector reforms has taken the center stage in the world's economy. Both developed and developing countries have tried to bring about reforms in their financial sectors in order to impact on the growth of either the entire economy or a sub-sector of the economy such as agriculture [1]. Financial sector reforms constitute that aspect of economic reforms which focuses mainly on restructuring the financial institutions (regulators and operators) via institutional and policy reforms [2]. It has also been described as deliberate measures and policies made by the relevant authorities to bring the needed changes in the financial institutions over a period of time [1]. These changes are expected to ultimately result in economic growth. Economic growth is defined in economic literature as the incremental outcome of productive activities, which can be assessed by observing the behavior of real Gross Domestic Product (GDP) or Gross National Product (GNP) per capita, year by year [3].

The financial system has been acknowledged to play an important role in economic growth and development [4,5]. Several theoretical and empirical studies at the international, national and provincial levels demonstrate that the financial sector could be a catalyst of economic growth if it is developed and healthy [6,7,8,9]. [10] opines that economic growth is significantly related to growth in agriculture in India, though it has declined in the recent years. Other researchers have lent credence to the fact that agriculture has important linkages and interrelate with the rest of the economy due to macroeconomic policies. [11] on his work on macroeconomic environment and agricultural sector growth in Nigeria stated that macroeconomic policies have directly and indirectly influenced agricultural output growth. He also reiterated that macroeconomic environment and other policies are not only used to regulate production activities in agricultural sector but in the other sectors of the economy. This interaction is highly vulnerable to changes in other sector especially macroeconomic policies not specially targeted at agriculture [12]. However, these macroeconomic policy outcomes vary greatly depending on the policy targets and

instruments used [13]. While the earlier works of Binswanger [14,15] support that agricultural production marketing and financing decisions are influenced by the macroeconomic environment. [16] also reported that in most other developing countries where agriculture is a large sector of the economy, no other sector of the economy is large enough to serve as an engine of economic growth in the next decade. This is because a large proportion of the Gross Domestic Product (GDP) comes from the agricultural sector as in the case of Nigeria.

According to [17], in Sub-Saharan Africa (SSA), the international recessions, debt crises and political instability adversely affected savings and investment ratio in the region. This was occasioned by a decline in the domestic resource mobilization and narrow tax base, which further depressed investment and economic growth. The unfavourable conditions in SSA relating to the deterioration in economic performance alongside financial repression in the 1970s gave rise to the adoption of structural adjustment programme under the auspices of the Washington Consensus [18]. In Nigeria, these unfavourable conditions resulted to financial sector reforms as a subset of the Structural Adjustment Programme in August, 1987 with interest rates being deregulated [19,20]. The key objective of the financial sector reforms was to create strong financial institutions that would take advantage of the benefits of increase in size, improve the efficiency and raise the diversity of the financial system of the economy. This objective was also to ensure that bank as financial intermediaries can contribute effectively to the agricultural sector through sound allocation of resources [21].

However, various governments in Nigeria over the years have initiated and implemented a myriad of financial reform measures, agricultural policies and programmes in an attempt to stimulate the sustainable growth and development of agricultural sector. Such policies include fiscal policies (like institutional creation and investment), exchange rate, pricing, and monetary policies [22]. Others that involve direct agricultural production through parastatals were River Basin Development Authorities (RBDAS), Directorate of food, Roads and Rural

Infrastructure (DFRRI) and the Agricultural Development Project (ADP) while those done through programmes were the National Accelerated Food Production Programme (NAFPP) of 1972, the Operation Feed the Nation (OFN) of 1976 and the Green Revolution (GR) of 1980 [23,20]. These policy measures were aimed at improving the sector to serve as the engine growth for other sectors [24]. In spite of several reform measures, there is still a knowledge gap regarding financial sector reforms and agricultural growth because research effort in this regard have been minimal, when compared to efforts the other components of the economic reforms such as trade liberalization and exchange rate reforms. Even where research is available, emphasis has tended to be placed on the institutional aspects of the reform that is banking sub-sector [25,2,26]. Mentions are only on the potential effects of the reforms on agriculture with no empirical evidence with efforts geared towards the investigation of current account and government deficits as well as their implication for saving and growth imbalances. Apart from this, many similar studies have failed to or not sufficiently document empirically the effect of reforms on agricultural growth [2,5,27]. It is against this backdrop that it becomes necessary to assess the impact of financial sector reforms on agricultural growth in Nigeria. The specific objective is to examine the effect of financial sector reforms on agricultural output growth in Nigeria.

2. LITERATURE REVIEW

The importance of the financial system to economic development is not a clear-cut issue. Researchers like [28] are of the view that economic development creates demand for certain financial instruments while others like [29] holds a contrary view and argues that the financial system plays a crucial role in the mobilization of capital for industrialization. Thus, the financial system only responds to the demand created as a result of economic development. It is however known that countries with better developed financial systems, that is financial markets and institutions with most effective way of channeling society's savings to its most productive use, tend to experience faster economic growth compared to those with less developed financial systems [30]. [31] submitted that institutions have direct and indirect benefits on economic growth and development. [32] on the relationship between institutions, macroeconomic policy and the growth of the

agricultural sector in Nigeria finds significant evidence in support of the hypothesis that institutions matter in economic growth especially the growth of the agricultural sector in Nigeria. This is because financial sector development helps economic growth through more efficient resource allocation and productivity growth rather than through the scale of investment or saving mobilization.

Although there is no single reform path, historical experience indicates that real and financial sector reforms can spur productivity growth and that is why the benefits of Productivity resulting from change in the component of output toward high-productivity sectors have played a vital role in some emerging market and developing economies [33]. They also stated that productivity growth in the tradable sectors (industry and agriculture) in emerging market economies exceeded that in services sector during the past decade of reforms while low-income countries were found to experience more significant productivity growth in the agriculture and service sectors. This explains why the growth and development of agricultural sector is fundamental for the overall process of socioeconomic development as various governments and institutions in the sub-Saharan African (SSA) sought for strategies that would lead to higher levels of production and a pivot factor for sustained increase of agricultural production in the improvement of productivity, which is carried out through technological and efficiency changes [34,35].

In agriculture, the physical inputs commonly used are land, labour, capital, management and water resource; part of this technological and efficiency changes are the reforms in the financial sector for effective and efficient mobilization of funds in the agricultural sector. For instance, [33] alluded that productivity-enhancing structural reforms were needed to boost technological catch-up, facilitate structural transformation into higher productivity sectors and new activities, and better allocate existing resources in the economy. According to [36], there are many areas in which reforms could have significant productivity impacts, either in the near-term or over the longer term. This was made possible by reforms in the financial sector which could be a catalyst of economic growth if it is developed and healthy [6,8]. Thus, the proper and timely reforms policies in the financial sector would enhance investment and growth in any sector such as agriculture since finance is postulated as important determinant of investment which culminates in growth [20].

Quite a number of empirical studies have been carried out to examine the effect of financial reforms on economic growth. The results confirm that there is a positive correlation between the two. [37] in his work on institutional reforms, interest rate policy and the financing of the agricultural sector in Nigeria using cointegration and an error correction mechanism (ECM) technique with annual time series data covering the period 1980 to 2011 posited that there is a negative relationship between agricultural value added, interest rate spread and inflation in the country. [38] examine the impact of financial sector reforms on agricultural and manufacturing sectors in Nigeria using the VAR methodology. The results indicate that bank credit to the private sector as a ratio of GDP has a positive effect on manufacturing and agricultural sectors in the short run, medium term and long term. Also, [9] in their works on The Impact of Financial Sector Reforms on the Nigerian Agricultural Export Performance using cointegration and error correction model (ECM) revealed that financial sector reforms significantly affect major agricultural export commodities such as cocoa, palm kernel and palm oil in Nigeria both in the long and short-run While other studies have increasingly found financial development to have a causal effect in stimulating economic and productivity growth [39,40,41,42,43].

Using long-term time-series data and the vector autoregressive (VAR) method of analysis, this study attempts to empirically examine the effect of financial sector reforms on the long-term agricultural growth in Nigeria. In addition to the stated objectives, it builds on the existing literature in two important ways. First, financial sector GDP or RGDP (value added) is used as a measure of financial sector reforms. This is a major departure from the commonly used bank related measures such as: monetary aggregates like M2, liquid liabilities of the financial system, or bank credit to the private sector which are regarded as poor indicators of financial sector reforms [44,45,46] because of: (i) they measure more the extent of monetization rather than financial sector reform, especially for the developing economies (ii) make no differentiation of liabilities among financial institutions;(iii) cannot represent the actual volume of funds channeled to the productive sector [47,48] as compared to financial sector Gross Domestic Product (GDP) or Real Gross Domestic Product

(RGDP) which is by far a better indicator of financial sector reforms [49,50] in a number of ways; (i) it represents a broader measure of financial reforms. (ii) it reflects all the activities of a financial system; that is, all financial transactions "involving the creation, liquidation, or change in ownership of financial assets and/or facilitating financial transactions".(iii) does not vary from structural changes within the financial sector.(iv) it does not underestimate the level of financial sector in Nigeria's economy, where a significant financial development, Investment, Productivity and economic growth or innovation occurs in the real sector. (iv) it uses a Granger causality testing procedure to conduct causality analysis.

2.1 Theoretical Framework

Many earlier studies have attempted to explain the interrelationship between financial sector reforms or financial development and economic growth using the endogenous growth theory, which shows that growth rates can be related to institutional arrangements [51,52,53,54]. However, little or no studies have been made to find the link between financial sector reforms and agricultural growth especially in developing country like Nigeria. In this study, it will adopt the earlier works of [55,50] on simple endogenous growth model, the "AK model", where aggregate real output is a function of the aggregate capital stock is used to illustrate the potential impacts of financial sector reform on agricultural growth.

$$Y_t = AK_t \qquad (1)$$

Where, Y_t and K_t are output and capital stock at time t, respectively and A is a constant measuring the amount of output produced for each unit of capital. Assuming, that a fraction of income, σ, is saved and invested, and dropping the time indices, the capital accumulation (investment) equation is given by:

$$\Delta K = \sigma Y - \delta K \qquad (2)$$

Where δ is the depreciation rate and both σ and δ are assumed to remain constant. Dividing both side of equation (2) by K results in the capital accumulation equation rewritten as: $\Delta K/K = \sigma Y/K - \delta$. Since, from equation (1), $Y/K = A$, substituting A for Y/K results in:

$$\Delta K/K = \sigma A - \delta \qquad (3)$$

Finally, by taking logarithms and derivatives of equation (1) and combining it with equation (3), the steady state growth rate can be written as:

$$y = \sigma A - \delta \qquad (4)$$

Where, y represent growth rate of output. Equation (4) indicates that the growth rate in output is the product of the saving rate and the marginal productivity of capital. Furthermore, it shows two ways through which financial sector reforms can affect agricultural growth. First, it increases σ, the saving rate, and thus, the investment rate. Second, it can increase A, the efficiency with which capital is used. The former effect is strongly emphasized by [56,57]. In the McKinnon-Shaw model, a well-developed financial system mobilizes savings by channeling the small-denomination savings into profitable large-scale investments. These savings might not be available for investment without the participation of financial institutions because mobilizing savings of disparate savers is usually costly due to the existence of information asymmetries and transaction costs. Financial institutions lower the cost of mobilizing savings and also provide attractive instruments and saving vehicles while offering savers a high degree of liquidity. According to [50], several theoretical models that emphasize the second, that is, the efficiency-enhancing and role of financial sector reform. These models show that financial sector reforms can affect productivity of capital in two major ways. (i) by collecting and processing information needed to evaluate alternative investment projects, thus improving the allocation of resources; (ii) by providing opportunities to investors to diversify and hedge risks, thereby inducing individuals to invest in riskier but more productive investment alternatives. These models affirmed that there is a positive two way causal relationship between financial sector reforms and agricultural growth thus showing that economic growth reduces the importance of fixed costs associated (incurred) in joining the financial market thereby facilitating the creation and expansion of more financial institutions.

3. MATERIALS AND METHODS

3.1 Data Collection

Secondary data for the study covering the period 1970-2009 were sourced from publication of the Central Bank of Nigeria Statistical Bulletin, 2009; Annual Report and Statements of Account of Central Bank of Nigeria (CBN) of various years,

National Bureau of Statistics (NBS), 2009 and National Population Commission of various years.

3.2 Analytical Techniques

The study made use of an econometric model adopted by [20] to express: (i) Agricultural investments as dependent on financial sector reforms (ii) Agricultural growth as dependent on financial sector reforms. The relationships were specified as follows:

$$AGINV = f(\text{FSRGDP}) \qquad (5)$$

$$AGRGDP1 = f(\text{FSRGDP}) \qquad (6)$$

Where:

$AGINV$ = Agricultural Investments (represented by Foreign Investment plus Domestic Investment. The Foreign Investment was proxied for Foreign Private Investment (FPI) in the agricultural sector while Domestic Investment was proxied for Credit to agriculture).

$AGRGDP1$ = Agricultural Growth (proxy for Growth Rate of Agricultural Sector Real Gross Domestic Product).

$LNFSRGDP$ = log of Financial Sector Real Gross Domestic Product (represented financial sector reforms)

3.3 Variables Description

SAV= Total Savings obtained from publication of Central Bank Nigeria (CBN, 2009) Statistical Bulletin as a measure economic activity; PSC = Private Sector Credit selected from financial Deepening Indicators (monetary aggregate) in CBN (2009) Statistical Bulletin to represents financial sector reforms in the sensitivity analysis; AGRGDP1 = Agricultural Growth (proxy for Growth Rate of Agricultural Sector Real Gross Domestic Product) was calculated from RGDP of agricultural sector from CBN (2009) Statistical Bulletin; FSRGDP = Financial sector reforms proxy for Financial sector RGDP representing RGDP of financial institutions sourced from CBN (2009) Statistical Bulletin; AGINV=Agricultural Investments (represented by Foreign Investment plus Domestic Investment. The Foreign Investment was proxied for Foreign Private Investment (FPI) in the agricultural sector while Domestic Investment was proxied for Credit to agriculture) obtained from publication of CBN (2009) Statistical Bulletin; ER = Exchange rate;

IR = Interest rate from CBN (2009) Statistical Bulletin publication; LFA = Labour Force in Agriculture (proxied by Agricultural labour force in the federal ministry) obtained from National Bureau of statistics (NBS) and National Population Commission (NPC) of various years; PCI = Per Capita income (calculated by dividing the GDP/Population) using data from Central Bank of Nigeria (2009) and NPC of various years.

From economic theory, other policy variables such as savings, income, output, interest rate and exchange rate also affect agricultural investments while agricultural investment, labour in agriculture, exchange rate and interest rate also affect agricultural growth. Therefore, we have:

$$LNAGINV = f(LNFSRGDP, LNSAV, LNAGRGDP1, LNPCI, LNER, LNIR) \quad (7)$$

$$LNAGRGDP1 = f(LNFSRGDP, LNAGINV, LNLFA, LNER, LNIR) \quad (8)$$

$LNSAV$ = Log of Total Savings
$LNPCI$ = Log of Per Capita Income
$LNLFA$ = Log of Labour Force in Agriculture
$LNER$ = log of Exchange Rate
$LNIR$ = log of Interest Rate

$(LNAGINV, LNAGRGDP1, LNFSRGDP)$ are logs of Agricultural Investments, Agricultural Growth and Financial Sector Reforms.

Given the various theories on the relationship between financial sector reforms and economic growth, various variables of interest such as

$(LNAGINV, LNAGRGDP1, LNFSRGDP, LNSAV, LNPCI, LNLFA, LNER, LNIR)$ were jointly determined. The empirical investigation into the relationships among these variables was carried out in a vector autoregressive (VAR) model and Granger causality test. A unique advantage of the VAR technique of analysis is that it treats all variables as potentially endogenous and also

facilitates investigation of the related concept of causality in the Granger's sense of it [58]. Causality in Granger's sense is inferred when values of a variable say X_t have explanatory power in a regression of Y_t on lagged values of Y_t and X_t. The vector autoregression (VAR) model has become one of the leading approaches employed in the analysis of dynamic economic relationships [59,60,61,62,20] like the ones specified in equations 7 and 8 respectively. This study follows suit by specifying a VAR model that examined the short and long-run relationship of the impact of financial sector reforms on agricultural investment and growth in Nigeria.

The VAR representation of the model with lag order k is thus:

$$Y_t = C_0 + \sum_{i=1}^{k} A_i Y_{t-i} + \mu_t \quad (9)$$

Where:

$Y_t = (LNAGINV, LNAGRGDP1, LNFSRGDP, LNSAV, LNPCI, LNLFA, LNER, LNIR)$ is a 8X1 vector of endogenous variables or Integrated Variables (10)

$C_0 = (C_1, C_2 \ldots C_n)$ the C intercept vector of the VAR model.

A_i = matrix coefficients estimated of autoregressive coefficient vector Y_{t-i}, for $i = 1, 2 \ldots k$. Thus, A_i is 8 x 8 coefficient matrices.

μ_t = $\mu_t = (\mu_{1t}, \mu_{2t}, \ldots \mu_{nt})$ vector of independent and identically distributed error terms (I.I.D).

k = the number of lagged terms.

The VAR estimations are very sensitive to structure of lag variables and sufficient lag length does help to reflect the long term impact of variables on others. However, including longer lag lengths will lead to multicollinearity problems and will increase the degrees of freedom (DOF) [63]. From equation (10), it was expanded as follows:

$$\Delta LNAGINV = \varphi_0 + \varphi_1 \sum_{i=1}^{k} \Delta LNAGRGDP1_{q,t-i} + \varphi_2 \sum_{i=1}^{k} \Delta LNFSRGDP_{q,t-i} + \varphi_3 \sum_{i=1}^{k} \Delta LNSAV_{q,t-i} + \varphi_4 \sum_{i=1}^{k} \Delta LNPCI_{q,t-i} + \varphi_5 \sum_{i=1}^{k} \Delta LNLFA_{q,t-i} + \varphi_6 \sum_{i=1}^{k} \Delta LNER_{q,t-i} + \varphi_7 \sum_{i=1}^{k} \Delta LNIR_{q,t-i} + \mu_{1t} \quad (11)$$

$$\Delta LNAGRGDP1 = \theta_0 + \theta_1 \sum_{i=1}^{k} \Delta LNAGINV_{q,t-i} + \theta_2 \sum_{i=1}^{k} \Delta LNFSRGDP_{q,t-i} + \theta_3 \sum_{i=1}^{k} \Delta LNSAV_{q,t-i} +$$
$$\theta_4 \sum_{i=1}^{k} \Delta LNPCI_{q,t-i} + \theta_5 \sum_{i=1}^{k} \Delta LNLFA_{q,t-i} + \theta_6 \sum_{i=1}^{k} \Delta LNER_{q,t-i} + \theta_7 \sum_{i=1}^{k} \Delta LNIR_{q,t-i} +$$
$$\mu_{2t} \qquad\qquad (12)$$

$$\Delta LNFSRGDP = \lambda_0 + \lambda_1 \sum_{i=1}^{k} \Delta LNAGINV_{q,t-i} + \lambda_2 \sum_{i=1}^{k} \Delta LNAGRGDP1_{q,t-i} + \lambda_3 \sum_{i=1}^{k} \Delta LNSAV_{q,t-i} +$$
$$\lambda_4 \sum_{i=1}^{k} \Delta LNPCI_{q,t-i} + \lambda_5 \sum_{i=1}^{k} \Delta LNLFA_{q,t-i} + \lambda_6 \sum_{i=1}^{k} \Delta LNER_{q,t-i} + \lambda_7 \sum_{i=1}^{k} \Delta LNIR_{q,t-i} +$$
$$\mu_{3t} \qquad\qquad (13)$$

$$\Delta LNSAV = \beta_0 + \beta_1 \sum_{i=1}^{k} \Delta LNAGINV_{q,t-i} + \beta_2 \sum_{i=1}^{k} \Delta LNAGRGDP1_{q,t-i} + \beta_3 \sum_{i=1}^{k} \Delta LNFSRGDP_{q,t-i} +$$
$$\beta_4 \sum_{i=1}^{k} \Delta LNPCI_{q,t-i} + \beta_5 \sum_{i=1}^{k} \Delta LNLFA_{q,t-i} + \beta_6 \sum_{i=1}^{k} \Delta LNER_{q,t-i} + \beta_7 \sum_{i=1}^{k} \Delta LNIR_{q,t-i} +$$
$$\mu_{4t} \qquad\qquad (14)$$

$$\Delta LNPCI = \gamma_0 + \gamma_1 \sum_{i=1}^{k} \Delta LNAGINV_{q,t-i} + \gamma_2 \sum_{i=1}^{k} \Delta LNAGRGDP1_{q,t-i} + \gamma_3 \sum_{i=1}^{k} \Delta LNFSRGDP_{q,t-i} +$$
$$\gamma_4 \sum_{i=1}^{k} \Delta LNSAV_{q,t-i} + \gamma_5 \sum_{i=1}^{k} \Delta LNLFA_{q,t-i} + \gamma_6 \sum_{i=1}^{k} \Delta LNER_{q,t-i} + \gamma_7 \sum_{i=1}^{k} \Delta LNIR_{q,t-i} +$$
$$\mu_{5t} \qquad\qquad (15)$$

$$\Delta LNLFA = \psi_0 + \psi_1 \sum_{i=1}^{k} \Delta LNAGINV_{q,t-i} + \psi_2 \sum_{i=1}^{k} \Delta LNAGRGDP1_{q,t-i} + \psi_3 \sum_{i=1}^{k} \Delta LNFSRGDP_{q,t-i} +$$
$$\psi_4 \sum_{i=1}^{k} \Delta LNSAV_{q,t-i} + \psi_5 \sum_{i=1}^{k} \Delta LNPCI_{q,t-i} \ \psi_6 \sum_{i=1}^{k} \Delta LNER_{q,t-i} + \psi_7 \sum_{i=1}^{k} \Delta LNIR_{q,t-i} +$$
$$\mu_{6t} \qquad\qquad (16)$$

$$\Delta LNER = \sigma_0 + \psi_1 \sum_{i=1}^{k} \Delta LNAGINV_{q,t-i} + \sigma_2 \sum_{i=1}^{k} \Delta LNAGRGDP1_{q,t-i} + \sigma_3 \sum_{i=1}^{k} \Delta LNFSRGDP_{q,t-i} +$$
$$\sigma_4 \sum_{i=1}^{k} \Delta LNSAV_{q,t-i} + \sigma_5 \sum_{i=1}^{k} \Delta LNPCI_{q,t-i} + \sigma_6 \sum_{i=1}^{k} \Delta LNLFA_{q,t-i} + \sigma_7 \sum_{i=1}^{k} \Delta LNIR_{q,t-i} +$$
$$\mu_{7t} \qquad\qquad (17)$$

$$]\Delta LNIR = \phi_0 + \phi_1 \sum_{i=1}^{k} \Delta LNAGINV_{q,t-i} + \phi_2 \sum_{i=1}^{k} \Delta LNAGRGDP1_{q,t-i} + \phi_3 \sum_{i=1}^{k} \Delta LNFSRGDP_{q,t-i} +$$
$$\phi_4 \sum_{i=1}^{k} \Delta LNSAV_{q,t-i} + \phi_5 \sum_{i=1}^{k} \Delta LNPCI_{q,t-i} + \phi_6 \sum_{i=1}^{k} \Delta LNLFA_{q,t-i} + \phi_7 \sum_{i=1}^{k} \Delta LNER_{q,t-i} +$$
$$\mu_{8t} \qquad\qquad (18)$$

While it is easier measuring other variables described above, it is also of valued importance to note that measuring financial sector reforms often poses a challenge to researchers in their effort to assess the impact of financial intermediation on real economic activity. The reason is as earlier stated above in introduction. Based on the above assertions, financial sector RGDP (Real Gross Domestic Product) was utilized as indicator for financial sector reforms.

The *apriori* expectations for vector autoregressive models are suited to track and identify structural shocks within a system of equations, with respect to underlying economic theory. The focus of this study is on the relationship between the impacts of financial sector reforms on agricultural investment and growth in Nigeria's economy, thus, we concentrate on the expected theoretical relationships that should hold in equation (13). The parameter for λ_1 is expected to be positively related. Financial constraint is one of the problems in agricultural production; an efficient financial sector should ensure the channeling of

funds to agricultural sector. Thus, as the financial sector becomes more efficient, more saving will be mobilized and this provides an opportunity for funds to be extended to provide investment opportunities in the agricultural sub-sector.

The parameter for λ_2 is expected to have a direct positive influence on income, interest rate, financial sector gross domestic product, financial sector reform and agricultural output growth. This implies that investment from the financial sector would greatly enhance agricultural output growth. The parameter for λ_3 is expected to be positively related. Generally, financial sector reform is expected to have a positive impact on saving mobilization in the economy. However, depending on the measure of the financial sector reform, a priori sign could be negative. The parameter λ_4 is expected to have a positive sign. Per capita income is very crucial for economic growth and it may increase savings, which may in turn help in boosting the financial system. The parameter λ_5 is expected to be positively related. The positive sign is to reflect the quality of manpower being produced by the agricultural sector of the economy.

The parameter λ_6 is expected to have a positive sign. This is because a lower interest rate will induce private economic agents to undertake investment activities at the lower levels of interest rate. However, in an environment characterized by severe financial repression as being the case in developing countries over the years, investment funds may not be readily available to potential investors [64]. In this case, the only way to induce people to mobilize investment funds through saving is high interest rate. This implies that the higher the financial intermediaries, the more the availability of investment funds through savings and hence the high level of investment in agriculture. This is the premise of the argument of Mckinnon-Shaw hypothesis which postulated a positive relationship between financial liberalization and the real interest rate. The parameter λ_7 is expected to have a negative sign since the exchange rate is negatively related to the agricultural production. At a favourable exchange rate, more agricultural outputs would be produced. Also, with good macroeconomic policies that enhance favourable exchange rates, agricultural funds can be widely available at low interest rate.

According to [63], VAR technique would be invalid if variables are not stationary at level, in such situation, a cointegration and vector error correction (VECM) techniques are carried out to investigate the relationship among non-stationary variables. Hence, it became necessary to conduct preliminary diagnostics on the time series properties of the variables before further evaluation. In order to ascertain the pattern of integration of the variables, a unit-root test was conducted using two specifications of the augmented Dickey-Fuller (ADF) test: (i) intercept (ii) trend and intercept. The latter was used for confirmation tests. The essence of the test was to determine stationarity in trend of the variable and to show the order of integration at which they become stationary if it reveals a non-stationary trend.

The hypothesis for the unit root test is:

o H_0: $\infty = 1$
o H_1: $\infty < 1$

To ensure the authenticity of the results obtained [65], cointegration test was carried out. This was done using the Johansen approach of testing the number of cointegrating vectors: the Trace and the Maximum Eigenvalue statistics. The null hypothesis for the trace test was that there are at most r cointegrating vectors, while for the Max Eigenvalue test, the null of $r = 0$ was tested against the alternative that $r = 1$; $r = 1$ was tested against the alternative that $r = 2$ and so on. The optimal lag length for the cointegration test was selected using the Schwarz Information Criterion (SIC).

After estimating the cointegtrated VAR, innovation accounting was conducted to determine the dynamic responses of the variables to one-standard deviation shocks in other variables in the system. This was done by generating the impulse response functions from the system. Impulse Response Functions (IRF), trace the responsiveness of the dependent variable in the VAR (VECM) to a unit shock in the error terms. For each variable from each equation, a unit shock was applied in the error term and the effects upon the VAR (VECM) to a unit shock in error terms are observed over a period of time. If there are K endogenous variables in the model, then a total of K^2 impulse responses can be generated. In this study, the analysis was confined to the responses of AGINV, AGRGDP, FSRGDP, SAV, PCI, LA, IR and ER to the shocks in FSRGDP.

To further obtain information concerning the relative importance of each innovation towards explaining the behaviour of the endogenous variables, variance decomposition analysis (VDC) was conducted. The generalized forecast error variance decomposition technique attributed by [66,67] were used. This technique has the advantage that its results are not sensitive to the ordering of the variables in the VAR (VECM).

To examine the short-run impacts of financial sector reforms on agricultural investments and growth in Nigeria, the Granger-casualty test developed by [58] was adopted. This test seeks to ascertain whether or not the inclusion of past values of a variable say Y_{t-p} do or do not help in the prediction of present values of another variable X. If X is better predicted by including past values of Y, than by not including them, then Y is said to Granger-cause X [68]. Several alternative methods of testing for Causality in Cointegrated VAR have emerged in the literature. The popular approach has been to re-parameterize the model into the equivalent vector error correction model (VECM) and to conduct Causality tests following either the residual-based Engle-Granger two-stage method or the Johansen-Type Error.

Finally, to ensure that the conclusions arrived at from the baseline models (equations 11 to 18) are not spurious or outcomes of chance, sensitivity analysis and robustness checks on the results was carried out. This was done by replacing a variable in the baseline model. The basic difference between the models used for sensitivity analysis and the baseline model was the introduction of Private Sector Credit (PSC) as a measure (or proxy) for Financial Sector Reforms in sensitivity analysis and robustness analysis while Financial Sector Gross Domestic Product (FSGDP) was used as a proxy for Financial Sector Reforms in the baseline equation.

The reduced form representation of the VAR is thus:

$$X_t = C_0 + \sum_{i=1}^{k} \delta_i X_{t-i} + \mu_t \qquad (19)$$

Where:

X_t
$= (LNAGINV, LNAGRGDP1, LNPSC, LNSAV, LNPCI, LNLFA, LNER, LNIR)$ is a 8x1 vector of five endogenous variables, while X_{t-i} is the corresponding lag term for each of the variables. A_i, is the 8 x 8 matrix of autoregressive coefficient vector X_{t-i}, for i = 1, 2,...k.
$C_0 = (C_1, C_2 \ ... \ C_n)$ is the C intercept vector of the VAR model.

$\mu_t = (\mu_{1t}, \mu_{2t,} \ ... \ \mu_{nt})$ is the 8x1 vector of independent and identically distributed error terms (I.I.D). K = the number of lagged terms.

LNPSC =   Log of Private Sector Credit to Agriculture

$(LNAGINV, LNAGRGDP1, LNSAV, LNPCI, LNLFA, LNER, LNIR)$ = are as earlier defined.

4. RESULTS AND DISCUSSION

4.1 Effect of Financial Sector Reforms on Agricultural Output

The result of the Augmented Dicker-Fuller (ADF) unit root test is presented in Table 1. Schwarz Information Criterion (SIC) was used for the selection of the optimal lag length to a maximum of 9. The results revealed that variables used in the analysis possessed unit-roots at one percent level of significance and was stationary only after transformation at the first differences for both intercept and when trend specification was included. Only the agricultural output growth variable was stationary at level, that is, I(0). With the above result, the unit-root test results gave a useful clue on how to arrange the variables into the vector error correction model (VECM) analysis. Thus, agricultural investments (LNAGINV), financial sector reforms (LNFSRGDP), total savings (LNSAV), per capita income (LNPCI), the labour force in agriculture (LNLFA), exchange rate (LNER) and interest rate (LNIR) were fed into the model at their first-differences, while agricultural growth (LNAGRGDP1) enters at its level.

With the fact that almost all the variables were stationary at the first differencing, it was necessary to carry out another test to assess if the non-stationary variables were co-integrated. In essence, the hypotheses were tested to affirm the rank of the cointegrating relationships that

Table 1. Result of ADF Unit root test

Variable	Levels		1st Difference		Conclusion
	Intercept	Trend + Intercept	Intercept	Trend + Intercept	
LNAGINV	-1.701[0]	-1.157[0]	-5.890[0]***	-6.161[0]***	I(1)
LNAGRGDP1	-6.039[0]***	-5.972[0]***	____	____	I(0)
LNFSRGDP	-2.819[0]	-2.392[0]	-5.935[0]***	-4.157[0]***	I(1)
LNSAV	1.130[0]	-0.771[0]	-4.661[0]***	-4.735[0]***	I(1)
LNPCI	-2.492[0]	-1.962[0]	-5.702[0]***	-6.046[0]***	I(1)
LNLFA	0.613[0]	-1.371[0]	-5.650[0]***	-5.856[0]***	I(1)
LNER	0.125[0]	-2.163[0]	-5.023[0]***	-4.995[0]***	I(1)
LNIR	-2.189[0]	-2.929[0]	-9.239[0]***	-9.162[0]***	I(1)

*Source: Computed by Author. Notes: *** indicates significance at 1% level. The values in bracket [ ] for the ADF test shows the optimal lag length selected by the SIC within a maximum lag of 9 .Variables are in log forms*

existed among the variables. Tables 2 and 3 show the results of Johansen cointegration tests indicating the presence of two (Trace) and one (Maximum Eigenvalue) cointegrating vectors respectively. This indicates that there were evidence of the existence of a long-run relationship among financial sector reforms, agricultural investment, output growth and other policy variables in Nigeria. Therefore, applying the vector error correction model (VECM) would enable us to track the long-run relationship between the variables and tie it to deviation that may occur in the short-run [69].

Table 2. Johansen cointegration trace test

Null hypothesis	Alternative hypothesis	Test statistic	Critical value 0.05
r = 0	r = < 1	198.682	159.530***
r = 1	r = < 2	134.750	125.615**
r = 2	r = < 3	91.215	95.754
r = 3	r = < 4	62.088	69.819
r = 4	r = < 5	38.263	47.856
r = 5	r = < 6	23.603	29.797
r = 6	r = < 7	10.765	15.494
r = 7	r = < 8	1.447	3.842

*Source: Computed by Author. Notes: r indicates the number of co-integrating vector. *** and ** are the significance levels at 1% and 5% respectively. P-values are obtained using response surfaces in [70]*

Table 3. Johansen cointegration maximum eigenvalue test

Null Hypothesis	Alternative hypothesis	Test statistic	Critical value 0.05
r = 0	r = 1	63.932	52.363***
r = 1	r = 2	43.535	46.231
r = 2	r = 3	29.128	40.078
r = 3	r = 4	23.825	33.877
r = 4	r = 5	14.659	27.584
r = 5	r = 6	12.839	21.132
r = 6	r = 7	9.318	14.265
r = 7	r = 8	1.447	3.842

*Source: Computed by Author. Notes: r indicates the number of co-integrating vector. *** and ** are the significance levels at 1% and 5% respectively. P-values are obtained using response surfaces in [70]*

Going by the Johansen cointegration results, a VECM (2) with at least two cointegrating vectors was carried out to ascertain that the estimated VECM was not false, the residual auto correlation and correlogram tests were also conducted. The results revealed that the residuals of the estimated VECM were appropriately uncorrelated, implying that the estimated VECM was correctly specified or

unbiased and the parameters estimated were consistent. This was because the spikes from the correlograms revealed the relative correlation of the error terms in the VECM equations and the closer the spikes are to the zero line, the more uncorrelated the error terms. The coefficients from the estimated VECM were not of primary interest in this empirical work. Instead, focused was on the impulse response function (IRFs) and variance decomposition (VDC) generated from the VECM.

The impulse response functions traced out the responsiveness of the dependent variable in the VECM to shocks on each of the variables using the Cholesky one standard deviation innovations (Choleskey one Standard deviations examine the dynamic interactions among variables). This implies that impulse responses showed the path of LNAGRGDP1 (agricultural growth) when there were innovations in the financial policy variables. For each equation, a unit shock was applied to the error, and the effects upon the VECM system over 30 periods were examined. The VECM system has eight variables, thus a total of 64 impulses could be generated. But the primary objective was to examine the impact of financial sector reforms shocks on the other seven macroeconomic or endogenous variables. Thus, only the responsiveness of the financial sector reforms on the macroeconomic variables (LNAGINV, LNAGRGDP1, LNSAV, LNPCI, LNLFA, LNER and LNIR) was traced out.

Fig. 1, presented seven panels of impulse response graphs indicating how innovations in financial sector reforms variable affected agricultural investments; growth and other policy variables in Nigeria over a period of 30 periods. Each panel illustrated the response of the policy variables to a one standard deviation innovation (corresponding to a positive shock) in the policy variable. A value of zero means that financial sector reforms shock has no effect on the financial policy variables and the variables continued on the same path it would have followed had there been no financial sector reforms shock. A positive or negative value indicated that the shock caused the variable to be above or below its 'natural' path as the case may be. The solid lines depict the estimated effects of the shock.

Panels A, B, C and G displayed the impulse responses of agricultural investments, agricultural growth, total savings and interest rate respectively to the one time shock in the financial sector reforms as presented in Fig. 1. These four financial policy variables fell below equilibrium with a negative response in an undulating but significant manner before stabilizing over the period reviewed to the positive shock in the financial sector reforms. This implies that there is a negative response among agricultural investments, savings and interest rate to the positive shock in financial sector reforms in Nigeria financial system. This would initially retard growth in agricultural sector as shown in panel B in the first two periods. However, in the long-run, it settles at a period below equilibrium level. That is, it leads to a long-run disequilibrium solution since the impulse response function does not return to the zero line. This is in line with the economic postulation that the higher the interest rate, the higher the savings and the greater the investment opportunities and vice versa. However, this was not consistent with our *apriori* expectations since economic agents are expected to adjust their spending and investment habits moderately and gradually in response to the increased supply of funds rather than immediately. The result correlated with that of [71] which reveals that low interest ceiling is seen to have restricted the real flow of loan-able funds, which depressed the quality of productive investment during financial sector reforms while [56] and [57] believe that when financial system is repressed by low level of savings rather than by the lack of investment opportunities, economic growth is severely hindered. Notwithstanding, investment is negatively linked to the effective real rate of interest on loans, but positively linked to the growth rate of the economy. They also opined that savings in many developing countries were barely sufficient to maintain the existing capital stock hence, could not permit enough investment to sustain economic growth.

The negative but significant effect of the financial sector reforms shock on these monetary variables displayed in Fig. 1 was not consistent with theoretical expectations and this confirms the weak and unstable nature of credit markets in the Nigeria economy but agree with [55] who stated that the financial sector reforms can affect economic growth by altering the saving rates. In this case, the sign of the relationship was not obvious, because financial sector development may also reduce savings and thereby growth while [72] affirms that the use of interest rate ceilings in a repressed system, distorts the economy in four critical ways: (i) Current consumption is favoured compared to future consumption (ii) financial institutions are

favoured instead of lending via deposits and engage potential investors in relatively low-yielding investments (iii) the low level of interest rates would cause borrowers to favour capital-intensive projects (iv) the pool of potential borrowers is dominated by entrepreneurs who possess low- yielding projects.

Panel D indicated that per-capita income was consistently falling below equilibrium over the time period, though exhibited rising trend at the beginning of the period. This is plausible because of the argument that focuses on the negative effect of repression on the rate of per-capita income contrary to the McKinnon-Shaw premise. Financial researchers argued that the increase in the real interest rate may not necessarily lead to improved private savings. In very poor countries for instance, the level of income would be so low that households spend a very high proportion of their earnings on basic needs. In such a case, even with high real interest rates, the very little proportion of income would be saved. This implies that McKinnon–Shaw's proposition would therefore be more relevant in rich countries. In Panel E, labour force in agriculture increased consistently in a positive manner over time. This implies that labour force is skewed to the shock in the reform with a positive future expectation. This is consistent with theoretical postulation reflecting the quality of manpower being produced by the agricultural sector of the economy. Panel F revealed that exchange rate responded positively after a period of the thirty years. This shows that financial sector reforms have a positive impact on monetary policy (exchange rate) in Nigeria. This is contrary to the negative theoretical postulation considering the seasonal nature of agricultural production in a country with an open economy having many trading partners.

The Variance Decomposition Analysis (VDC) provides a means of analysis to determine the relative importance of the dependent variable in explaining the variations in the explanatory variables. The result of variance decomposition over a 30-year time period is displayed in Table 4. The values in the table confirmed the results obtained from the Impulse response analysis (IRFs). On the average, 3.84 percent of most of the variation in the forecast error for financial sector reforms was explained by the shocks to itself. Agricultural growth shock had a value of 78.84 percent of the variation in financial sector reforms over the 30-year period while the impact of financial sector reforms shock on agricultural investments was 0.34 percent over the 30 year period.

The influence of the shocks on total savings was marginal at the onset with values less than one percent up to the fifteen periods but later increased to 1.37 percent over the thirtieth year period. The average contributions of LNPCI, LNLFA, LNER and LNIR were 2.49 percent, 10.53 percent, 5.19 percent and 0.08 percent respectively. The higher contributions of agricultural growth to the shocks in financial sector reforms indicate that financial sector reforms significantly affected agricultural output than other monetary variables which were mild and persistent. This correlated with the empirical works of [73,44] which confirms that financial development has a positive effect on economic growth. The implication of this is that policy makers will be faced with less uncertainty in planning for the long-term.

However, comparing the VDC analysis for agricultural output growth in Tables 5 revealed that agricultural output shocks on financial sector reforms was lower with an average contribution of 33.35 percent respectively than the shocks of financial sector reforms on agricultural output with an average of 78.84 percent as shown in Table 4. This implies that an increase in the financial sector through effective reforms would enhance growth in the agriculture sector of the country. On the other hand, agricultural output shocks on other policy variables showed a positive impact indicating a positive interaction among variables.

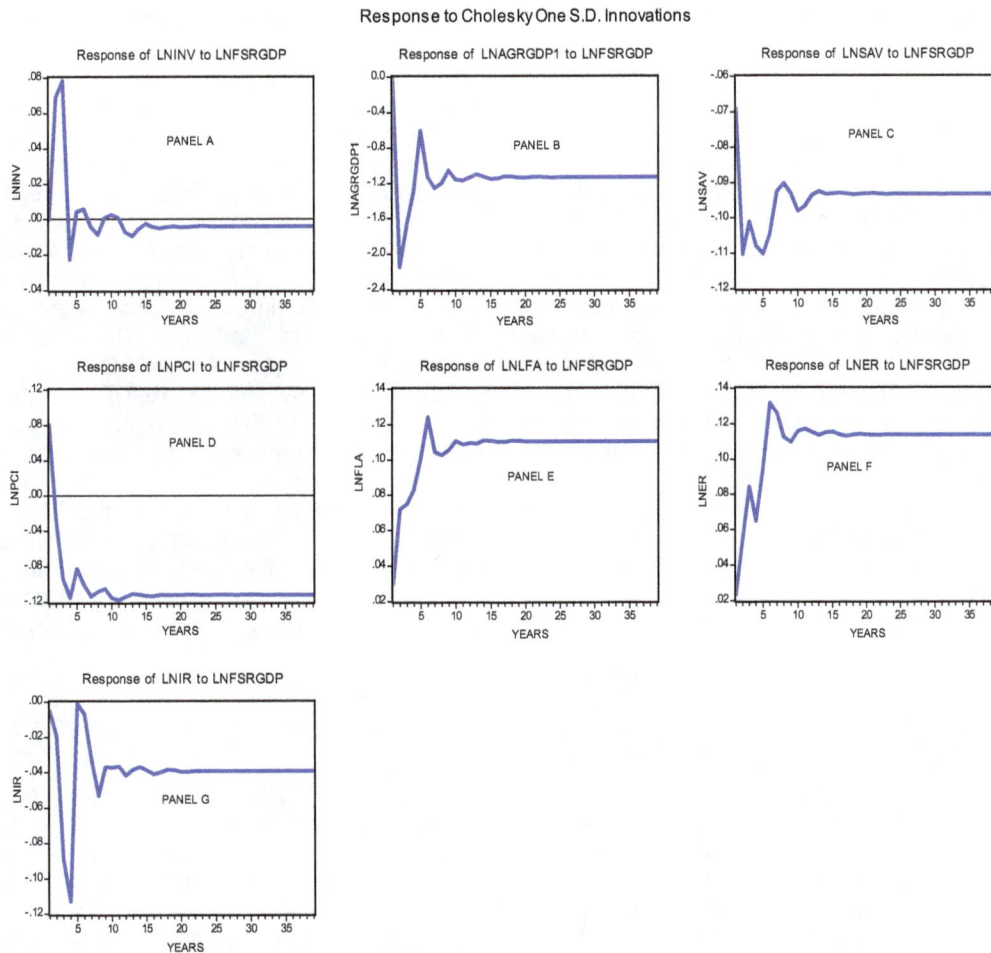

Fig. 1. Response of LNAGINV, LNAGRGDP1, LNSAV, LNPCI, LNLFA, LNER, and LNIR to LNFSRGDP shock in choleskey one standard deviation (Choleskey one Standard deviations examine the dynamic interactions among variables)

The low values recorded in interest rates are good signals to the agricultural investments which would consequently impact positively on output growth in agriculture as it would attract investors to come and invest into the sector.

To examine if a significant short-run relationship existed between financial sector reforms and the macroeconomic variables used in the study, an error correction modeling (ECM) analysis was employed as presented in Table 6. The parsimonious estimate showed that R^2 value of 0.97 indicates the variables explained about 97 percent of agricultural output growth. F-statistic of 31.62 (P<0.01) reveals that they are jointly significant and the Durbin Watson Statistic value of 2.05 implies that the model does not suffer from autocorrelation problem but has a very good fit. For significance variables, it was found out that financial sector reforms (LNFSRGDP), past value of financial sector reforms (proxy by LNFSRGDP), per capital income (LNPCI), labour force in agriculture (LNLFA) and interest rate (LNIR) were the significant determinants of agricultural output growth in Nigeria for the period of analysis. This result implies that the past financial sector reforms, per capital income, labour force in agriculture and interest rate significantly increases the current financial sector reforms, per capital income, labour force in agriculture and interest rate. Further result shows that the coefficients of other three variables (savings, exchange rate and agricultural investments) were not significantly different from zero. This suggests that past financial sector reforms, per capital income, labour force in agriculture and interest rate positively determine the current flow of financial sector reforms, per capital income, labour force in agriculture and interest rate in Nigeria while the previous savings, exchange rate and agricultural

investments do not significantly affect the present savings, exchange rate and agricultural investments in Nigeria. As such, an increase in the previous volume of financial sector reforms, per capital income, labour force in agriculture and interest rate would result in an increase in the present level of financial sector reforms, per capital income, labour force in agriculture and interest rate to agricultural output in Nigeria. The result further reveals that the coefficient of the error correction term which measures the speed of adjustment towards long-run equilibrium was both correctly signed (negative) and statistically significant at 1% percent which shows a yearly correction of about 75% of the error with a different adjustable speeds to long-run equilibrium. Thus, correcting any deviations from the long-run equilibrium. The implication of this is that financial sector reforms had an impact on agricultural growth in the short-run.

Finally, a sensitivity analysis/ robustness check was carried out on VECM equation (15) to ascertain that the model was not spurious or outcome of chance. The results indicated that private sector credit (LNPSC) which replaced financial sector real gross domestic product as a proxy for financial sector reforms was stationary at first difference in Augmented Dickey-Fuller (ADF) for both intercept and when trend specification was added. The Johansen cointegration results of LNPSC and other variables (LNINV, LNAGRGDP1, LNSAV, LNPCI, LNLFA, LNER, LNIR) earlier specified indicated the presence of two and one cointegrating vectors in the system at 1 percent significant level for both Trace and Maximum Eigenvalues vectors as displayed in Tables 7 and 8 respectively. This confirms the existence of a long-run relationship between financial sector reforms and the policy variables in Nigeria.

Table 4. Variance decomposition of FSRGDP

Period	S.E	LNINV	LNAGRGDP1	LNFSRGDP	LNSAV	LNPCI	LNLFA	LNER	LNIR
1	0.396	88.336	10.223	0.000	0.000	0.000	0.000	0.000	0.000
5	0.725	0.582	81.492	5.078	0.300	0.702	6.676	5.108	0.062
10	0.946	0.426	79.893	4.080	0.599	1.821	9.341	3.763	0.077
15	1.124	0.341	78.846	3.838	0.951	2.493	10.530	2.917	0.083
20	1.279	0.286	78.328	3.688	1.163	2.839	11.169	2.443	0.086
25	1.417	0.251	78.022	3.585	1.289	3.047	11.567	2.151	0.089
30	1.542	0.226	77.812	3.515	1.374	3.189	11.841	1.952	0.091

Source: Computed by Author. Note: S.E (Standard Error), LNINV (log of Agric. Investment), LNAGRGDP1 (log of Agric. Growth), LNFSRGDP (Log of Financial Sector RGDP), LNSAV (log of total savings), LNPCI (Log of Per Capita Income), LNLFA (Log of Labour Force in Agric.), LNER (Log of Exchange Rate) and LNIR (log of Interest Rate)

Table 5. Variance decomposition of LNAGRGDP1

Period	S.E	LNINV	LNAGRGDP1	LNFSRGDP	LNSAV	LNPCI	LNLFA	LNER	LNIR
1	5.028	0.052	99.948	0.000	0.000	0.000	0.000	0.000	0.000
5	6.481	7.000	66.001	22.378	0.660	2.078	2.941	1.910	0.032
10	7.418	9.499	50.747	29.250	1.546	2.529	3.132	3.268	0.028
15	8.211	11.074	43.062	33.350	2.083	2.822	3.308	4.265	0.026
20	8.925	12.140	37.871	36.214	2.415	3.014	3.445	4.877	0.024
25	9.586	12.928	34.084	38.303	2.653	3.155	3.542	5.314	0.022
30	10.204	13.532	31.183	39.901	2.835	3.263	3.615	5.649	0.021

Source: Computed by Author. Note: S.E (Standard Error), LNINV (log of Agric. Investment), LNAGRGDP1 (log of Agric. Growth), LNFSRGDP (Log of Financial Sector RGDP), LNSAV (log of total savings), LNPCI (Log of Per Capita Income), LNLFA (Log of Labour Force in Agric.), LNER (Log of Exchange Rate) and LNIR (log of Interest Rate)

Table 6. Vector error correction model for short-run impact

Variable	Coefficient	Std. Error	t-Statistic	Prob.
C	-1.695538	0.842285	-2.013022**	0.0613
D(LNAGRGDP1(-2))	0.160707	0.099081	1.621968	0.1243
D(LNFSRGDP)	9.746098	1.774025	5.493778***	0.0000
D(LNFSRGDP(-2))	-5.517377	1.671753	-3.300354***	0.0045
D(LNSAV)	3.561411	2.476003	1.438371	0.1696
D(LNSAV(-1))	1.763468	2.343236	0.752578	0.4626
D(LNPCI)	2.854077	1.655964	1.723514	0.1041
D(LNPCI(-1))	-5.845329	1.326259	-4.407382***	0.0004
D(LNPCI(-2))	1.599045	1.586716	1.007770	0.3286
D(LNLFA)	2.912682	1.724201	1.689294	0.1106
D(LNLFA(-2))	-2.028902	1.127346	-1.799715*	0.0908
D(LNER)	-1.863357	1.486112	-1.253846	0.2279
D(LNER(-1))	-1.678422	1.090944	-1.538505	0.1435
D(LNIR)	-1.206719	1.174977	-1.027015	0.3197
D(LNIR(-1))	-1.982491	1.023264	-1.937418*	0.0706
D(LNIR(-2))	-0.642033	0.764057	-0.840295	0.4131
D(LNINV)	0.828225	0.687123	1.205352	0.2456
D(LNINV(-1))	1.108967	0.642909	1.724920	0.1038
D(LNINV(-2))	1.006459	0.688385	1.462058	0.1631
ECM(-1)	-0.751859	0.140136	-5.365209***	0.0001
R-squared	0.974060	Mean dependent var		0.024444
Adjusted R-squared	0.943257	S.D. dependent var		5.895698
S.E. of regression	1.404400	Akaike info criterion		3.817279
Sum squared resid	31.55744	Schwarz criterion		4.697011
Log likelihood	-48.71101	Hannan-Quinn criter.		4.124329
F-statistic	31.62198	Durbin-Watson stat		2.050879
Prob(F-statistic)	0.000000			

*Source: Computed by Author. Note: *** ** and * =1%, 5% and 10% significance levels respectively.*

The impulse response function analysis in robustness test showed that variables responded in a similar pattern to the one time shock of financial sector reforms (LNPSC) as that of the baseline model (LNFSRGDP). This could probably be due to the order of integration of LNPSC which was stationary at first difference at 1 percent significance as earlier reported. The

Variance Decomposition Analysis (VDC) results revealed that the average values of 5.41 percent, 0.60 percent, 2.50 percent, 0.67 percent, 5.66 percent, 1.09 percent, 0.30 percent of LNAGINV, LNAGRGDP1, LNSAV, LNPCI, LNLFA, LNER and LNIR respectively contributed to the variations in financial sector reforms (83.82 percent) over the 30-years. The impact of financial sector reforms shock (proxy by LNPSC) on agricultural output growth was lower with 0.60 percent when compared with 78.85 percent of Table 4 (baseline Table). This is because a well structured, effective and efficient financial sector reforms is expected to impact more on agricultural growth at a lower value of 3.84 percent which is the case of financial sector reforms (LNFSRGDP) in the baseline model than the higher value of 83.82 percent LNPSC obtained in sensitivity analysis with agricultural growth value of 0.60 percent. This implies that the financial sector reform was not strong, effective and efficient to cause an astronomical growth in the agricultural sector. This result is in line with several theoretical and the empirical studies of [6, 7] at the international, national and provincial levels who affirmed that the financial sector could be a catalyst of economic growth if it is well developed and healthy. The result further agrees with the theoretical argument and is consistent with the results of VAR model; Cross-country/panel empirical studies of [50,74,75] which provided evidence that financial sector real gross domestic product used as proxy in financial sector reforms or financial development has greater significance and positive effect on long-term growth through its investment and productivity effects than proxies used in monetary aggregates such as private sector credit (PSC). This implies that financial sector reforms played a significant role in the growth of agricultural sector by increasing its productivity level and independently generated positive investments in the sector which was a converse of the sensitivity result (PSC). A similar trend was experienced for per capita income (0.67 percent), labour force in agriculture (5.57 percent) and exchange rate (1.09 percent) were lower in sensitivity test except for savings (2.46 percent) and interest rates (0.30 percent) that were higher in sensitivity test than that of the baseline test.

Furthermore, the shock of LNAGRGDP1 on LNPSC was lower in sensitivity analysis with 14.25 percent than 33.35 percent shocks on LNFSRGDP recorded in Tables 5 (baseline table) respectively. From the above results, it could be deduced that financial sector reforms

shocks greatly influenced microeconomic policy variables in the baseline model especially agricultural output growth than those financial sector reforms shocks in sensitivity analysis in the long-run during the period under review. In the short-run, the coefficient of the error correction term was statistically significant and correctly signed (negative) indicating a different adjustable speed to long-run equilibrium. Thus, correcting any deviations from the long-run equilibrium. This affirms that financial sector reforms positively impacted on agricultural growth in the short-run just as in the base line analysis.

Table 7. Johansen cointegration trace test

Null hypothesis	Alternative hypothesis	Test statistic	Critical value 0.05
r = 0	r = < 1	216.539	159.530***
r = 1	r = < 2	130.879	125.615**
r = 2	r = < 3	86.672	95.754
r = 3	r = < 4	55.859	69.819
r = 4	r = < 5	34.548	47.856
r = 5	r = < 6	20.901	29.797
r = 6	r = < 7	9.524	15.494
r = 7	r = < 8	0.773	3.842

Source: Computed by Author. Notes: r indicates the number of co-integrating vector. *** and ** are the significance levels at 1% and 5% respectively. P-values are obtained using response surfaces in [70]

Table 8. Johansen cointegration maximum eigenvalue test

Null hypothesis	Alternative hypothesis	Test statistic	Critical value 0.05
r = 0	r = 1	85.660	52.363***
r = 1	r = 2	44.207	46.231
r = 2	r = 3	30.813	40.078
r = 3	r = 4	21.311	33.877
r = 4	r = 5	13.647	27.584
r = 5	r = 6	11.376	21.132
r = 6	r = 7	8.752	14.265
r = 7	r = 8	0.773	3.842

Source: Computed by Author. Notes: r indicates the number of co-integrating vector. *** is the significance level at 1%. P-values are obtained using response surfaces in [70]

5. CONCLUSION AND RECOMMENDATION

The vector error correction model (VECM) result revealed that financial sector reforms had significant effects on agricultural output growth in

Nigeria both in the long and short-run in the baseline model and the sensitivity analysis/robustness check respectively. Also, it was discovered that financial sector reforms in the baseline test impacted more on microeconomic policy variables especially growth in agricultural output than those in the sensitivity test in the long-run. Based on the outcome of this work, it is concluded that the assertions made by [50,46,49,76] regarding the use of financial sector real gross domestic product as a better indicator of financial sector reforms than monetary aggregates such as private sector credit commonly used as a proxy for financial sector reforms is true since it impacted more positively on agricultural investments and growth in Nigeria. It is therefore suggested that the Government adopts strong macroeconomic policies targeted to bring meaningful growth in the agricultural and financial sector against foreign-based economic policies since the results confirmed the significant impact of financial sector reforms on agricultural growth in Nigeria in the long and short-run.

COMPETING INTERESTS

Author has declared that no competing interests exist.

REFERENCES

1. Akpaeti AJ. Impact of financial sector reforms on agricultural investments and growth in Nigeria (1970-2009). Unpublished Ph.D Dissertation, Department of Agricultural Economics, Michael Okpara University of Agriculture, Umudike, Nigeria. 2012;1-197.
2. Balogun ED. A review of Soludo`s perspective of banking sector reforms in Nigeria. MPRA Paper No. 3803. 2007;(3-5, 7-8, 12-14).
3. Eyo EO. Agricultural development in nigeria plans, policies and programmes. best print business press, Uyo, Nigeria. 2005;1.
4. Babalola RM. Remark by Mr. Babalola honorable minister of state for finance at the may, 2008 members evening round table talks held at the IOD National Secretariat, Ikoyi, Lagos on May; 2008.
5. Okorie G, Uwaleke UJ. An Overview of Financial Sector Reforms and Intermediation in Nigeria.
6. Bullion. 2010;34(2):19-29.

7. Chen BZ, Feng Y. Determinants of economic growth in china: private enterprise, education and openess. China Economic Review. 2000;11:1-15.
8. Chow GC, Li KW. China's Economic Growth: 1952-2010. Economic Development and Cultural Change. 2002; 51: 247-256.
9. Daramola AS. Competitiveness of Nigerian Agriculture in a Global Economy: any Dividend(s) of Democracy?' Inaugural Lecture Series 36, delivered at the Federal University of Technology, Akure, Nigeria; 2004.
10. Akpaeti AJ, Bassey NE, Ibok OW. The impact of financial sector reforms on the nigerian agricultural export performance. American Journal of Experimental Agriculture. 2014;4(9):1072-1085.
11. Marjit S. Financial sector reform for stimulating investment and economic growth–The Indian Experience. Policy paper prepared for the Ministry of Finance, Government of India and the ADB, New Delhi, 1; 2004.
12. Eyo EO. Macroeconomic environment and agricultural sector growth in Nigeria. World Journal of Agricultural Science. 2008;4(6): 781-786.
13. Oyejide TA. The effects of trade and exchange rate policies on agriculture in Nigeria. Research Report 55. International Food Policy Research Institute; 1986.
14. Agu C. What does the Central Bank of Nigeria Target? An analysis of monetary policy reaction function in Nigeria. Final report submitted to the african economic research consortium, Nairobi, Kenja, December; 2007.
15. Binswanger H. The policy response of agriculture. In: Stanley F. & De Tray. F (Eds.). Proceedings of the World Bank Conference on Development Economics. Washington DC; 1989:121-153.
16. Kwanashie M, Garba A, Ajilima I. Policy Modeling in Agriculture: testing the response agriculture to adjustment policies in Nigeria. AERC Research Paper, 57; 1997.
17. World Bank (1989). From Crisis to Sustainable Growth. A Long-term Perspective Study, Washington DC, Kasekende LA, Atingi-Ego M. financial liberation and its implications for the domestic financial system: The case of Uganda. African Economic Research

Consortium (AERC) Paper 128, Nairobi, February, 1; 2003.

18. World Bank. Uganda moving beyond recovery: investment and behavior change, for growth country economic memorandum. Report No.39221-UG; 2007.

19. Ikhide SI. Financial sector reforms and monetary policy in Nigeria. IDs Working Paper 68. 1996;1-4.

20. Akpaeti AJ. Does financial sector reforms affect agricultural investments in Nigeria? A Cointegration and VAR Approach. International Journal of Food and Agricultural Economics. 2013;1(2):13-28.

21. Central Bank of Nigeria. Fifty Years of Central Banking in Nigeria 1958-2008. 135,136,172,200,201; 2009.

22. Ukoha OO. Relative price variability and inflation: evidence from the agricultural sector in Nigeria. African Economic Research Consortium, Nairobi. AERC Research paper . 2007;171:1-21.

23. Ekpo HA, Ndebbio JEU, Akpakpan EB, Nyong MM. Macro-Economic Model of the Nigerian Economy Vantage. Publishers Ibadan, Nigeria. 2004;3-7.

24. Akinboye OL. Five decades of agricultural policies in Nigeria: What Role has Statistics Played? Bullion. 2008:32(4):35-44.

25. Ikhide SI, Alawode AA. Financial sector reforms macroeconomic instability and the order of the economic liberialization: The Evidence from Nigeria African Economic Research Consortium (AERC) Final Report; 1994.

26. Anyanwu CM. An overview of current banking sector reforms and the real sector of the nigerian economy. Central Bank of Nigeria Economic and Financial Review. 2010;48(4):31-56.

27. Ugwu DS, Kanu IO. Effects of agricultural reforms on the agricultural sector in Nigeria. Journal of African Studies and Development. 2012;4(2): 51-59.

28. Robinson J. The Generalization of the General Theory. In: The Rate of Interest and other Essays. London: Macmillian; 1952.

29. Hicks JR. A Theory of Economic History. Clarendon Press, Oxford; 1969.

30. Beck T, Levine R, Loayza N. Financial and Sources of Growth. Journal of Financial Economics. 2000;58:261-300.

31. Adebiyi M.A. Industrial finance in Nigeria: performance, problems and prospects In: industrialization, urbanization and development in Nigeria, 1950-1999, Edited by: MOA. Adejugbe, Concept Publications. 2004;408-428.

32. Omojimite BU. Institutions, macroeconomic policy and the growth of the agricultural sector in Nigeria. Global Journal of Human Social Science. 2012;12(1):1-9.

33. Dabla-Norris E, Ho G, Kochhar K, Kyobe A, Tchaidzel R. anchoring growth: The importance of productivity-enhancing reforms in emerging market and developing economies. international monetary fund strategy, policy, and Review Department. 2013;1-36.

34. Ajao AO. Empirical analysis of agricultural productivity growth in Sub-Sahara Africa: 1961 – 2003. Libyan Agricultural Resource Center Journal International. 2011;2(5):224 - 231.

35. Akpaeti AJ, Umoh GS. Farm resource productivity in conflict communities: evidence from the niger delta region, Nigeria, Sky Journal of Agricultural Research. 2013;2(3):28 – 39.

36. Hausmann R, Rodrik D, Velasco A. Growth Diagnostics. In: The Washington consensus reconsidered: towards a new global governance, ed. by J. Stiglitz and N. Serra, Oxford University Press, New York; 2008.

37. Kolawole BO. Institutional reforms, interest rate policy and the financing of the agricultural sector in Nigeria. European Scientific Journal. 2013;9(12): 259-272.

38. 38. Udah EB, Obafemi FN. The impact of financial sector reforms on agricultural and manufacturing sectors in Nigeria: An Empirical Investigation. Journal of Sustainable Development in Africa. 2011; 13(8):193-213.

39. Moore T, Green JC, Murinde V. Financial sector reforms and stochastic policy simulations: A flow of funds model for India. Journal of policy Modeling. 2006; 28(3):319-333.

40. Das A, Ghosh S. Financial deregulation and efficiency: An empirical analysis of indian banks during the post reform period.' In Review of Financial Economics. 2006;15 (3):193-221.

41. Limi A. Banking sector reforms in pakistan: economies of scale, scope and cost complimentarities. Journal of Asian Economics. 2004;15,3(6):507-528.

42. Spiegel MM, Yamori N. The impact of Japan's Financial Stabilization Laws on

bank equity values. In Journal of the Japanese and International Economies; 2003;17(3): 263-282.

43. Honda E. Financial deregulation in Japan. In Japan and the World Economy. 2003; 15(1):135-140.

44. Levine R, Loayza N, Beck T. Financial intermediation and growth: Causality and causes. Journal of Monetary Economics. 2000;46:31-77.

45. Arestis P, Luintel K.Financial Development and Growth: The role of stock markets. Journal of Money, Credit and Banking. 2001; 30:16-41.

46. Liang Q, Teng JZ. Financial development and economic growth. China Economic Review. 2006;17:395-911.

47. Demetriades PO, Husswein KA. Does Financial development cause economic growth? Time-series evidence from 16 Countries. Journal of Development Economics. 1996;51: 387-411.

48. Luintel KB, Khan M. A quantitative reassessment of the Finance-Growth Nexus: Evidence from a Multi-Variate VAR. Journal of Development Economics. 1999; 60:381-405.

49. Watchel O. Growth and Finance: What do we know and how do we know it? International Finance. 2000;4:335-356.

50. Ghirmay T. Financial development, investment, productivity and economic growth in the U.S. Southwestern Economic Review. 2009;23-29.

51. Romer P. Increasing Returns and long-run growth. Journal of Political Economy. 1986;94:1002-37.

52. Lucas R. On the mechanics of economic development. Journal of Monetary Economics.1988;22:3-42.

53. Barro R. Economic growth in a cross section of countries. Quarterly Journal of economics. 1991;106:407-443.

54. Grossman G, Helpman E. Innovation and Growth in the Global Economy, Cambridge: MIT Press; 1991

55. Asamoah GN. The Impact of the financial sector reforms on savings. Investment and growth of gross domestic products (GOP) in Ghana. International Business and Economic Research Journal, October. 2008;74.

56. McKinnon R. Money and Capital in Economic Development. Washington, D.C Brooking Institution; 1973.

57. Shaw ES. Financial Deepening in Economic Development. New York: Oxford, University. 1973;58.

58. Granger CWJ. Investigating causal relationships between econometric models and cross-spectral methods. Econometrica. 1969;37(3):424-438.

59. Barsky RB, Kilian L. Oil and the Macroeconomy since the 1970s. Journal of Economic Perspectives. 2004;18(4):115-134.

60. Kilian L. Not all oil price shocks are alike: disentangling demand and supply shocks in the crude oil market, American Economic Review. 2009;99:1053-1069.

61. Ozlale U, Pekkurnaz D. Oil Prices and Current account: a structural analysis for Turkish Economy, Energy Policy; 2010. DOI:10.1016/j.enpol.2010.03.082..

62. Chuku AC, Effiong EL, Sam NR. Oil Price Distortions and their Short and Long-run impacts on Nigerian Economy. Paper presented at the 51st annual conference of the Nigerian Economic Society (NES) held in Abuja, between 25th-27th October; 2010.

63. Wooldridge J. Introduction Econometrics: A Modern Approach. Thompson South-Western, New York; 2006.

64. Afangideh UJ. Financial development and in investment in Nigeria: Historical Simulation Approach. Journal and Monetary Integration. 2009;9(1):74-97.

65. Maddala GS. Introduction to econometrics. john wiley and sons, New Delhi India; 2002.

66. Koop G, Pesaran MH, Potter SM. Impulse response analysis in non-linear multivariate models. Journal of Econometrics. 1996;74(1):119-147.

67. Pesaran MH, Shin Y. Generalized Impulse response Analysis in Linear Multivariate Models, Economics Letters. 1998;58(1):17-29.

68. Gujarati D. Basic econometrics. Tata McGraw Hill, New Delhi; 2003.

69. Lorde T, Jackson M, Thomas C. The macroeconomic effects on a price fluctuation on a small open oil-producing country: The Case of Trinidad and Tobago. Energy Policy. 2009;37(7):2708-2716.

70. MacKinnon JG, Haug AA, Michelis L. Numerical distribution functions of likelihood ratio tests for cointegration. Journal of Applied Econometrics. 1999;14:563–77.

71. Hasan MA, Akhan, A, Ali SS. Financial sector reforms and its impact on

Investment and Economic Growth: An Economic Approach. Pakistan Development Review; 1996.

72. Fry M. Money and capital or financial deepening in economic development. Journal of Money, Credit and Banking, 1988;10(4):464-474.

73. King, RG, Levine R. Financial and Growth: Shumpeter Might Be Right. Quarterly Journal of Economics. 1993;108:717-737.

74. Levine R, Zervos S. Stock Market Development and Long-Run Growth. World Bank Economic Review. 1996;10: 323-339.

75. Benhabib J, Spiegel M. The role of Financial development in growth and investment, Journal of Economic Growth 2000;5:341-360.

76. Neusser K, Kugler M. Manufacturing growth and financial development: Evidence from OECD Countries. Review of Economics and Statistics. 1998;80:638.

8

Performance and Carcass Evaluation of Broilers fed Whole Millet Meal in a Humid Tropical Environment

L. A. F. Akinola[1*], O. A. Ekine[1] and C. C. Emedo[1]

[1]*Department of Animal Science and Fisheries, Faculty of Agriculture, University of Port Harcourt, P.M.B. 5323, Port Harcourt, Rivers State, Nigeria.*

Authors' contributions

This work was carried out in collaboration between all authors. Authors LAFA, OAE and CCE jointly designed the study, managed the collection of data and performed the statistical analysis. All authors read and approved the final manuscript.

<u>Editor(s):</u>
(1) Zhen-Yu Du , School of Life Science, East China Normal University, China.
(2) Anonymous.
<u>Reviewers:</u>
(1) Anonymous, Brazil.
(2) Anonymous, India.
(3) Anonymous, USA.

ABSTRACT

This study was conducted using 96 day old broiler chicks, to assess the most appropriate level of inclusion of whole millet meal as a substitute for maize. The inclusion levels of the whole millet were 0, 20, 40 and 60% of the maize content of the control diet. The birds were randomly assigned to four dietary treatments which were replicated four times using Completely Randomized Design (CRD). The study lasted for 56 days. Feed and water were provided *ad libitum* during the study. The weight of the birds, daily feed intake, feed cost per treatment and mortality records were kept. At the end of the experiment, one bird per replicate was randomly selected for carcass and organ evaluation. These were expressed as percentages of the body weight. Data collected were subjected to Statistical Analysis System (SAS) and errors were presented as standard errors of means (SEM). The inclusion of whole millet significantly influenced feed intake and feed conversion ratio. Significant differences were not observed in carcass, organ and other parameters measured. It was observed from this study that the performance, carcass and organ yield of broiler fed up to 60% whole millet in the diets as substitute for maize were equivalent to those fed the control diet. However, the 20% whole millet inclusion which gave the best feed conversion ratio and had similar final weight compared with the others was recommended for best broiler performance in a humid tropical environment.

Corresponding author: E-mail: letorn.akinola@uniport.edu.ng;

Keywords: Whole millet; broiler; carcass; organ; humid environment.

1. INTRODUCTION

Maize, a conventional dietary energy source is the most expensive ingredient in poultry feeds because of the quantity required for inclusion [1] which constitute about 50-70% of the total feed ingredients in broiler ration. Several attempts had therefore been made to reduce, maintain or regulate the cost of feed production by replacing part of the maize in diet with industrial by-products and others [2].

Pearl millet which is also called Gero or Maiwa or Dauro is a cereal with superior amino acid balance and higher protein content than corn [3,4,5]. It has higher oil content than other cereals [4,6] and a better source of linolenic acid. The findings of [7] exonerated millet from anti-nutritional properties (phytate and tannins) while [8] reported that millet has no tannin, contains 5-7% oil and has higher protein and mineral than maize. The concentration of anti-nutritional factors in pearl millet is generally lower than in other cereals and in general, anti-nutritional factors are not a problem with pearl millet [9]. The phytate content of pearl millet is similar to sorghum and maize in the range of 1.7-3.3 g/kg [10]. According to [11] millet contains 2984Kcal/kg metabolizable energy (ME), 11.55 crude protein (CP), 3.6% crude fibre (CF), 6.5% fat, 3.2% ash, 0.23% lysine and 0.15% methionine.

The use of whole pearl millet (up to 50%) in the temperate region gave equivalent or better results when the performance and carcass yield of broiler were evaluated [12,13]. This finding improved the feasibility of using pearl millet in poultry diet since it can be utilized successfully in whole form. It was therefore necessary to assess the suitability of using whole millet in broiler diet in the humid tropics (since the cost of grinding will be eliminated) with reference to performance, carcass and organ yield and feed cost.

2. MATERIALS AND METHODS

2.1 Experimental Location

The research was conducted at the Poultry Unit of the Teaching and Research Farm of Rivers State University of Science and Technology, Port Harcourt, Rivers State, Nigeria. Port Harcourt lies between longitude 6°55N to 7°10E and latitude

4°35N to 4°54N of Greenwich meridian, covering a total area of 804 km^2 [14].

2.2 Experimental Feed

The feed ingredients were obtained from the open market (Oyigbo market). The whole millet was incorporated into broiler starter and finisher diets at 0, 20, 40 and 60% in four treatments (T1-T4), to replace the maize content in the control (T1). The metabolizable energy (ME), crude protein (CP), fat and crude fibre (CF) of the diets were calculated while proximate analysis was conducted to obtain the analyzed values of CP, fat and CF using the method of [15]. All diets were isocaloric (2700 Kcal/kg and 3000 Kcal/kg) and isonitrogenous (20.60% and 18.20%) for starter and finisher diets respectively (Tables 1 and 2).

2.3 Experimental Birds

A total of 96 day old broilers were used for the study which lasted for 8 weeks. They were randomly assigned to the 4 treatments on arrival in a Completely Randomized Design (CRD). Each treatment had 4 replicates with 6 birds each (24 birds per treatment). Feed and water were provided *ad libitum* throughout the study period. Birds were fed the starter and finisher diets for 4 weeks each. Routine vaccination and management practices were duly carried out throughout the period.

2.4 Data Collection

The initial weight of the birds was taken on arrival while the final weight record was taken at the end of the experiment. The feed left over was subtracted from the quantity offered to determine the daily feed intake. The weight gain by the birds, feed conversion ratio (total feed consumed per bird/total weight gain per bird) and feed cost per treatment were computed while mortality record was taken daily. Data collection was done in accordance with the rules of the Ethics Committee of the Faculty for animal research.

2.5 Carcass and Organ Evaluation

At the end of the experiment, one bird was randomly selected per replicate for carcass and organ evaluation. They were fasted for 8 hours, bled, scalded, de-feathered, eviscerated and dissected for the determination of carcass and organ weights. Cut parts and organs were weighed on fresh basis using a sensitive scale (Metler balance) but were recorded as percentages of the life weight.

2.6 Data Analysis

All the data collected were subjected to analysis of variance using [16]. Significant differences between the means were separated using Duncan's Multiple Range Test as outlined by [17].

3. RESULTS AND DISCUSSION

The result of the performance of broilers fed the various levels of whole millet in diet is shown in Table 3. There were no significant (p > 0.05) differences between the birds fed the control diet and the whole millet diets in final weight, weight gain and mortality. This may be due to the uniformity in the nutrient content of diets. This finding was in line with [13] who reported that the performance and carcass yield of broilers fed diets containing whole millet were equivalent to

those fed a typical corn-soybean meal diet. The birds fed the 20% whole millet diet had significantly better total feed consumption per bird and feed conversion ratio. This was similar to the report by [12] who stated that the performance and carcass yield of broilers fed diets containing up to 50% ground pearl millet were equivalent or better than those fed typical corn-soybean meal diets. The use of 20% whole millet in diets for broilers in the humid tropics, thus seem to be of advantage from this study. The numerical value obtained from the cost of feed per bird in each treatment was similar even though the actual cost of maize was N85/kg and millet N110/kg. This similarity in cost of feed per bird across the treatment was attributed to the elimination of grinding cost for millet since the whole grain was used. The mortalities of birds were not significantly different in the treatments. The mortalities recorded occurred within the 2 days of arrival and could be attributed to the stress of transportation.

Table 1. Composition of experimental starter diet

Ingredients (%)	T_1 (0%WM)	T_2 (20% WM)	T_3 (40%WM)	T_4 60%WM)
Maize	44.0	35.2	26.4	17.6
Whole millet	0.0	8.8	17.6	26.4
Soya bean meal (44%CP)	20.0	18.0	18.0	18.0
Palm kernel cake	17.0	20.0	21.0	22.0
Fish meal (63%CP)	5.0	5.0	5.0	5.0
Wheat bran	9.9	8.9	7.9	6.9
V/tm premix	0.25	0.25	0.25	0.25
Bone meal	3.0	3.0	3.0	3.0
DL methionine	0.03	0.03	0.03	0.03
Salt	0.8	0.8	0.8	0.8
Lysine	0.02	0.02	0.02	0.02
Nutrient composition (%/kg)				
ME (Kcal/kg)	2721.52	2709.00	2702.15	2701.04
Crude protein (calculated)	20.61	20.60	20.62	20.69
Crude protein (analysed)	20.52	20.56	20.51	20.53
Fat (calculated)	3.57	3.60	3.61	3.61
Fat (analysed)	3.58	3.59	3.60	3.60
CF (calculated)	5.85	6.10	6.09	6.20
CF (analysed)	5.72	5.98	5.93	6.00
Lysine (calculated)	1.05	1.04	1.02	1.02
DL-Methionine(calculated)	0.41	0.42	0.42	0.43
Available phosphorus (calculated)	1.03	1.03	1.02	1.02

Vitamin and Trace Mineral Premix: Vitamin A, 551 IU; Vitamin D3, 110 IU; Vitamin E, 1.1 IU; Vitamin B12, 0.001 mg; riboflavin, 0.44 mg; niacin, 4.41 mg; D-panthotenic, 1.12 mg; choline, 19.13 mg; menadione sodium bisulfate, 0.33 mg; pyridoxine HCL, 0.47 mg; thiamine, 2.2 mg; D-biotin, 0.011 mg; and ethoxyquin, 12.5 mg. Mn, 6.0; Zn, 5.0: Fe, 3.0; I, 1.5; Se, 0.5

Table 2. Composition of experimental finisher diet

Ingredients (%)	T_1 (0% WM)	T_2 (20% WM)	T_3 (40% WM)	T_4 (60% WM)
Maize	52.8	42.2	31.68	21.12
Whole millet	0.0	10.6	21.12	31.68
Soya bean (44%)	14.2	13.2	11.0	10.0
Palm kernel (63%)	20.4	21.4	22.6	23.6
Fish meal (63%)	5.0	5.0	6.0	6.0
Palm oil	3.0	3.0	3.0	3.0
V/TM Premix	0.25	0.25	0.25	0.25
Bone meal	3.5	3.5	3.5	3.5
DL methionine	0.03	0.03	0.03	0.03
Salt	0.8	0.8	0.8	0.8
Lysine	0.02	0.02	0.02	0.02
Nutrient composition (%/kg)				
ME (KCal/kg)	3049.27	3044.24	3037.90	3032.11
Crude protein (calculated)	18.22	18.23	18.20	18.21
Crude protein (analysed)	18.20	18.21	18.17	18.19
Fat (calculated)	6.65	6.70	6.80	6.80
Fat (analysed)	6.63	6.68	6.72	6.73
CF (calculated)	5.30	5.77	5.90	5.90
CF (analysed)	5.31	5.72	5.81	5.82
Lysine (calculated)	0.91	0.90	0.89	0.88
DL-methionine (calculated)	0.38	0.38	0.39	0.39
Available phosphorus (calculated)	0.99	0.99	1.00	1.01

Vitamin and Trace Mineral premix: Vitamin A, 551 IU; Vitamin D3, 110 IU; Vitamin E, 1.1 IU; Vitamin B12, 0.001 mg; riboflavin, 0.44 mg; niacin, 4.41 mg; D-panthotenic, 1.12 mg; choline, 19.13 mg; menadione sodium bisulfate, 0.33 mg; pyridoxine HCL, 0.47 mg; thiamine, 2.2 mg; D-biotin, 0.011 mg; and ethoxyquin, 12.5 mg. Mn, 6.0; Zn, 5.0: Fe, 3.0; I, 1.5; Se, 0.5

Table 3. Effect of whole millet diets on broiler performance

Parameters	T1 (0% WM)	T2 (20% WM)	T3 (40% WM)	T4 (60% WM)	SEM
Initial weight (g/bird)	40.00	40.00	40.00	40.00	
Final weight (g/bird)	2080	2070	2070	2080	0.03
Weight gain (g/bird)	2040	2030	2030	2040	11.00
*TFC (g/bird/day)	4356.80[a]	3528.00[b]	4239.20[a]	4300.80[a]	72.50
Feed conversion ratio	2.14[a]	1.74[b]	2.09[a]	2.11[a]	0.35
Feed cost/bird (N)	388.77	385.62	383.10	376.24	
Mortality (%)	0.48	0.48	0.00	0.24	

*TFC = Total Feed Consumed, SEM = Standard Error of Mean, WM = Whole millet; *a,b – Means within each row that ear different superscript differ significantly; No. of replicates /treatment = 4

Tale 4. Effect of whole millet diets on carcass and organ yield of broilers

Parameters (%)	T1 (0% WM)	T2 (20% WM)	T3 (40% WM)	T4 (60% WM)	SEM
Breast	21.03	20.99	20.97	20.94	0.47
Back	17.32	17.01	17.00	16.99	0.45
Thigh	14.62	14.98	14.60	14.53	0.19
Neck	5.23	5.24	5.18	5.20	0.30
Shank	5.11	5.12	5.09	5.10	0.28
Wing	9.37	9.35	9.32	9.30	0.18
Heart	0.74	0.73	0.74	0.72	0.14
Liver	2.08	2.03	2.02	2.05	0.12
Gizzard	4.25	4.19	4.20	4.22	0.13

SEM – Standard Error of Mean; No. of replicates /treatment = 4

The carcass weight of birds expressed as the percentage of the live weight was not significantly different (p > 0.05) in the various treatments (Table 4 above). This may be as a result of the non-significant difference recorded in the final weight and weight gain of birds in all the treatments. The organ weight (heart, liver and gizzard) of birds recorded as percentage of the live weight had similar trend like the carcass weight. The gizzard weight of birds fed the whole millet diets which were not significantly different from birds fed the control diet was contrary to the finding of [18,19,20] who found increase in gizzard weight when whole grain such as wheat, triticale, millet and barley were fed to broilers. The similarity in the percentages of the gizzard in this study may either be attributed to the uniformity in the crude fibre (CF) content of the diets or to the small number of birds sampled in the current research.

4. CONCLUSION

The inclusion of 20 – 60% whole millet diet supported broiler production in the humid tropics as it gave equivalent result with those fed the control diet. But the 20% whole millet inclusion which gave better feed conversion confirmed the feasibility of using whole millet in poultry diet in this region. More importantly, the use of the whole millet in broiler diet which eliminated the cost of grinding as well as some extra work that could have been done during the grinding process will be of interest to farmers.

COMPETING INTERESTS

Authors have declared that no competing interests exist.

REFERENCES

1. Collins VP, Cantor AH, Pescatore AJ, Straw ML, Ford MJ. Evaluation of pearl millet as feed ingredient for poultry. Poult. Sci. 1994;73(Suppl.1):34. (Abstr.).
2. Ogbonna JU, Oredein AO, Ige AK. Effects of varying broiler levels of cassava leaf meal on broiler gut morphology. In Proc. of 25th Ann. Conf. NSAP. 2000;143–146.
3. Sullivan TW, Douglas JH, Andrews DJ, Bowland PL, Hancock JD, Bramel-Cox PJ, Stegmeier WD, Brethour JR. Nutritional value of pearl millet for food and feed. In Proc. Int'l. Conf. Sorghum Nutr. Qual. Purdue Univ. West Lafeyette. 1990;83-94.
4. Adeola O, Rogler JC. Pearl millet in diets of white pekin ducks. Poult. Sci. 1994;73:425-435.
5. Amato SV, Forrester RR. Evaluation of pearl millet as feed ingredient for broiler rations. First Natl. Grain Pearl Millet Symp. Tifton, G.A. and Teare, I.D. eds. Univ. of Georgia, Coastal Plain Expt. Station. 1995;125-128.
6. Hill GM, Hanna WW. Nutritive characteristics of pearl millet grain in beef cattle diet. J. Anim. Sci. 1990;68:2061-2066.
7. Cromwell GL, Coffey RD. An assessment of the availability of phosphorus in feed ingredients for non-ruminants. Proc. Maryland Nutr. Conf. for feed Manufacyurers, Maryland, USA. 1993;146-158.
8. NRC. National Research Council. Nutrient Requirement for Domestic Animals. 9th Edition. National Academy Press, Washington, D.C. USA; 1996.
9. Andrew DJ, Kumar KA. Pearl millet for food, feed and forage. Advances in Agronomy. 1992;48:89-139.
10. Rajashekher RA, Prasad VIK, Rao CIN, Sudhakar D. Sorghum and pearl millet for poultry feed. In Alternative uses of sorghum and pearl millet in Asia. Proceedings of the Expert Meeting, ICRISAT, Patancheru, A:P, India. CFC Technical Paper No. 33. 2003;264-277.
11. Aduku AO. Tropical Feedstuff Analysis Table. Department of Animal Science, Faculty of Agric. Ahmadu Bellow Univ. Zaria, Nigeria; 1993.
12. Davis AJ, Dale NM, Ferreira FJ. Pearl millet as an alternative feed ingredient in broiler diet. J. Appl. Poult. Res. 2003;12:137-144.
13. Hidalgo MA, Davis AJ, Dale NM, Dozier WA. Use of whole pearl millet in broiler diet. J. of Appl. Poult. Res. 2004;13:229-234.
14. Akpokodje GE. Street map of Port Harcourt and environ, Ibadan Kraft Books. 2001.
15. AOAC. Association of Official Analytical Chemist. Official Methods of Analysis. 15th ed Washington DC.
16. SAS Institute. SAS/STAT User's Guild. SAS Institute Inc., Cary, NC, USA; 1990.
17. Steel RGD, Torrie JH. Principles and procedures of statistics a biometrical approach. 2nd Edition, New York McGraw-Hill Book Company; 1980.

18.   Kiiskinen T. Feeding whole grain with pelleted diets to growing broiler chicken. Agric. Food Sci. Finland. 1996;5:167–175.

19.   Svihus B, Herstad O, Newman CW, Newman RK. Comparison of performance and intestinal characteristics of broiler chickens fed on diets containing whole, rolled or ground barley. Br. Poult. Sci. 1997;38:524-529.

20.   Jones GPD, Tylor RD. The incorporation of whole grain into pelleted broiler chicken diets: Production and physiological responses. Br. Poult. Sci. 2001;42:477-483.

9

Performance and Haematological Profiles of Crossbred Male Rabbits Fed Yam and Cassava by Products in the Humid Tropics

Joseph S. Ekpo[1*], Nseabasi N. Etim[1], Glory D. Eyo[1], Edem E. A. Offiong[1] and Metiabasi D. Udo[1]

[1]Department of Animal Science, Akwa Ibom State University, P.M.B.1167, Uyo, Akwa Ibom State, Nigeria.

Authors' contributions

This work was carried out in collaboration between all authors. Author JSE designed the study, wrote the protocol and wrote the first draft of the manuscript. Author NNE reviewed the experimental design and all drafts of the manuscript. Author GDE managed the analyses of the study. Author EEAO identified the plants. Author MDU performed the statistical analysis. All authors read and approved the final manuscript.

Editor(s):
(1) Zhen-Yu Du, School of Life Science, East China Normal University, China.
(2) Anonymous.
Reviewers:
(1) N. Nwachukwu, Dept. of Biochemistry, schl. of science, Fed. University of Technology, Owerri, Nigeria.
(2) Fernanda Carlini Cunha dos Santos, Universidade Federal de Pelotas, Pelotas, Rio Grande do Sul, Brazil.

ABSTRACT

This study was conducted to evaluate growth rate and haematological profiles of crossbred weaner males rabbits fed cassava and yam peels meal diets. Thirty-six crossbred weaner male rabbits aged 5-6 weeks were randomly allocated to 4 dietary treatments. Each treatment having 3 replicates with 3 rabbits per replicate in a completely randomized design. Diet 1 was composed by 37% of maize (control group), diet 2-37% of yam peel, diet 3-37% cassava peel and diet 4-37% yam-cassava peel mix. The experiment lasted for 12 weeks. Results obtained revealed that rabbits that received diets 2 performed better (P<0.05) than those that were fed diets 1, 3 and 4 in terms of daily weight gain. Haematological parameters assessed indicated no treatment effect (P>0.05) among the groups. It is concluded that yam peel, cassava peel, and yam – cassava peel mix could successfully replace maize in rabbits diets.

Corresponding author: E-mail: jsekpo@yahoo.com;

Keywords: Cassava peel; growth; haematology; rabbit; yam peel.

1. INTRODUCTION

Rabbits in the tropics are raised at ambient temperatures ranging between 27.44ºC and 28ºC which is outside their comfort zone of 21-23ºC [1]. Yet, rabbit production has been identified as one of the means of attaining sufficiency in the supply of animal protein to the diets of average Nigerians [2]. Rabbit can produce 6 pounds (2.72 kg) of meat on the same feed as a cow will produce 1 pound (0.45 kg) of meat on the same feed [3]. In addition rabbit meat is richer in protein (21%) compared with chicken (19.5%), beef (20%) and pork (17%) [4].

They have potential as meat-producing animals in the tropics. This make rabbit production a new impetus in Africa amongst a wide range of people there by creating the need for alternative cheap sources of rabbit feed to replace cereals in rabbit diets in order to make rabbit production profitable. Among such alternatives are cassava and yam peels which are readily available with little cost. Nigeria is the world largest producer of cassava and yam [5,6], producing more than 4 million tones of cassava and yam peels, annually as livestock feed [7]. These make the cost of these by-products 40% lower than that of maize [8]. A successful replacement of conventional feed ingredients should reduce feed cost in rabbit production. Rabbit production provides a rapid means of increasing animal protein supply because they are known to have such attributes as short generation interval, rapid growth rate and ability to utilize forage agro-industrial by-products and kitchen waste [2]. The present study was designed to evaluate growth rate and haematological profile of crossbreed weaner rabbits fed cassava and yam peel meal diets in the humidtropics.

2. MATERIALS AND METHODS

2.1 Site of Experiment

The experiment was conducted at the teaching and research farm of Akwa Ibom State University Obioakpa campus, Akwa Ibom State University Nigeria. Obioakpa is situated on latitude 5º 28^{1}N and longitude 7º 32^{1}E. It has mean annual rainfall of 2200 mm, annual humidity of 95%, with average temperature of 27ºC [9].

2.2 Source and Processing of Test Materials

The test feedstuffs (yam and cassava peels) which were collected as kitchen waste from the Akwa Ibom State University environs, were sundried for 4 days each before milling in a hammer mill to produce cassava peel meal (CPM), and yam peel meal (YPM). Four diets were formulated: diet 1 (control) contained 37.00% maize as the main energy source. The 37.00% maize in diet 1 was replaced with yam peel meal in diet 2, cassava peel meal in diet 3 and mixture of yam and cassava peel at equal ratio in diet 4. The experimental diets were supplemented with legume (centrosema). The ingredient composition of the diet is shown in Table 1. The diets supplied approximately 17.50% crude protein.

2.3 Experimental Animals

A total of thirty-six cross breed (Dutch x Newzealand) male rabbits, aged 5-6 weeks, an average weight of 582.00 g were used in the experiment. The rabbits were conditioned for two weeks before randomly allotted to the four dietary treatments in groups of 9 rabbits each. Each treatment group was replicated three times with three (3) rabbits per replicate (i.e. 9 rabbits per diet). Each individual rabbit was housed in a hutch measuring 70 cm x 40 cm. The entire hutch system was of the three-tier model. The rabbits were given experimental diets and clean water *ad libitum* throughout the experimental period of 12 weeks.

2.4 Data Collection

Feed samples were analyzed for proximate composition according to the procedure of AOAC [10]. The proximate composition of yam and cassava peels are presented in Table 2.

Daily feed consumption was obtained by subtracting the weight of the leftover feed each morning from the weight feed offered the morning of the previous day. Initial weight of the rabbits were taken the first day of the experiment and subsequently on weekly basis.

At the end of the 12 weeks feeding trial, a total of twenty-four rabbits (two per replicate) were randomly selected, starved overnight. Blood samples for haematological analysis was collected through marginal ear vein into bottles containing ethylene diamine tetracetic acid (EDTA). The blood samples were analyzed for packed cell volume (PCV), haemoglobin concentration (Hb), and white blood cell count (WBC) using the methods described by Monica [11]. The mean corpuscular haemoglobin concentration (MCHC), mean corpuscular haemoglobin (MCH) and mean corpuscular volume (MCV) were calculated.

2.5 Data Analysis

Data collected from the experiments were subjected to analysis of variance (ANOVA) [12]. Treatment means were compared using Duncan's New Multiple Range Test (DNMRT) where significant treatment effects were observed. Differences in means were considered statistically significant at p<0.05 [13].

3. RESULTS AND DISCUSSION

The performance of rabbits in the feeding trial is shown in Table 3. There were significant differences (P<0.05) among treatments in final body weight, daily weight gain and daily feed intake. Rabbits fed diet 2 had the highest final body weight (1798.95 g) while those fed diet 1

had the lowest final body weight (1622.87 g). Similarly, rabbits fed diet 2 had the highest daily weight gain (14.62 g) while the rabbits fed diet 3 had the lowest daily weight gain (12.35 g). The poor performance in weight gain of rabbits fed diet 3 may be attributed to the lowest dietary energy of diet 3 compared to other diets. This is because reduced dietary energy often limits growth in monogastrics including rabbits [14,15]. On the contrary, weight gain was significantly higher (P<0.05) in rabbits fed diet 2. This could be invariably linked to the relatively increased dietary energy in relation to diets 3 and 4. It also suggest that the higher protein level of diet 2 might have been well utilized by the animal compared to diet 1. [16,17] had earlier observed that increase dietary protein could contribute to increased growth of animals. Increase crude protein result in increase lean body mass (muscle), which indicate growth.

Feed intake value of 66.98 g, obtained from rabbits fed diet 3, was observed to be highest while the lowest value of 61.00 g was obtained from rabbits fed diet 1. The significant increase in feed intake of rabbits fed diet 3 may be attributed first to the low dietary energy content of diet 3 in relation to diets 1, 2 and 4. This is in agreement with [2] who observed that rabbits usually consume more of low energy feeds such as

Table 1. Composition of experimental rabbit diets

Ingredients	Treatments			
	T_1 (control)	T_2 (YPM)	T_3 (CPM)	T_4 (combined)
Maize	37.00	-	-	-
Yam peel	-	37.00	-	18.50
Cassava peel	-	-	37.00	18.50
Soybean meal	14.00	12.00	16.00	16.00
Wheat offal	30.00	30.00	27.00	28.50
Palm kernel cake	14.50	16.50	14.00	14.00
Fish meal	1.00	1.00	2.50	1.00
Bone meal	1.50	1.50	1.50	1.50
Oyster shell	1.50	1.50	1.50	1.50
Vit/premix*	0.25	0.25	0.25	0.25
Salt	0.25	0.25	0.25	0.25
Total	100	100	100	100
Calculated nutrient composition				
Crude protein	17.67	17.87	17.46	17.87
ME(Kcal/g)	2.52	2.35	2.06	2.23
Fibre	5.67	8.78	8.79	8.80
Ash	3.94	6.80	5.86	6.33
Ether	4.60	3.77	5.48	4.62

Each kg contained vit A. 8500 IU; vit D_3. 2000 IU; vit E, 8000 IU; vit K_3. 1.50 mg; vit B_1, 3.20 mg; vit B_6, 1.80 mg; vit B_{12}, 10 meg; pantothenic acid, 1.50 mg; folic acid, 0.50 mg; biotin, 0.20 mg; choline 0.20 mg; manganese, 0.75 mg; zinc,0. 45 mg; Iron 0.20 mg; Copper, 0.35 mg; Seleniun, 0.20 mg; Cobalt, 0.20 mg; Antioxidant 0.125 mg

Table 2. Proximate composition of yam peel and cassava peel

Parameter (%)	Yam peel	Cassava peel
Dry matter	89.90	91.24
Crude protein	11.14	5.19
Crude fibre	6.30	19.72
Ash	7.30	7.26
Ether extract	4.12	5.02
NFE	71.14	62.81

Table 3. Effects of experimental diets on performance of weaner rabbits (n=36)

Parameters	T_1(control)	T_2(yam peel)	T_3(cassava peel)	T_4(yam/cassava)
Initial body height	580.98	570.81	584.98	589.89
Final body weight	1,692.91^c	1,798.95^a	1,622.87^d	1,683.89^b
Total weight gain	1,111.93^b	1,228.14^a	1,037.89^d	1,094.00^{bc}
Daily weight gain	13.24^b	14.62^a	12.35^d	13.02^{bc}
Daly feed intake	61.00^c	63.00^b	66.98^a	62.50^{bc}
Feed conversion ratio	4.61	4.31	5.42	4.80

^{abc} means along the same row having no superscript are not significantly different (P>0.05)
SEM = standard error mean

Table 4. Effects of experimental diets on haematological indices of rabbits (n=36)

Parameter	Treatments			
	T_1 (control)	T_2 (YPM)	T_3 (CPM)	T_4 (YPM/CPM)
Packed cell volume (%)	37.42	38.50	36.80	37.40
Haemoglobin (g/dl)	11.45	12.16	10.70	11.40
Red blood cell (x10^6/mm^3)	5.68	5.75	4.50	5.20
White blood cell (x10^3/mm^3)	7.36	6.82	7.26	6.98
MCV (F1)	6.59	6.69	8.20	7.19
MCH (Pg)	20.16	21.15	23.78	21.92
MCHC (g/100 ml)	30.59	31.58	29.10	30.48

^{abc} means along the same row having no superscript are not significantly different (P>0.05)
SEM = standard error mean

cassava peel meal to satisfy their energy needs. The result of this study corroborate Olorunsanya et al. [18] who observed highest daily feed intake when 100% maize was replaced with sun-dried cassava peel meal.

Feed conversion values obtained indicated non-significant difference (P>0.05) among treatment although the highest value (5.42) was obtained from rabbits fed diet 3 and the lowest value (4.31) from rabbits fed diet 2. The similarity in feed conversion ratio of the rabbits on this trial is an indication that replacement of maize with cassava peel meal or yam peel did not impair nutrient utilization in the growing rabbits.

3.1 Haematological Indices

The results of the haematological indices are presented in Table 4 above. There was no significant difference (P>0.05) among the

treatment. Packed cell volume was highest (38.50%) in rabbits fed diet 2 while those fed diet 3 had the lowest (36.30%). Similarly, highest Hb concentration (12.16 g/dl) was obtained from rabbits fed diet 2 while the lowest Hb value (10.70 g/dl) was obtained from rabbits fed diet 3. There were no significant difference (P>0.05) among the treatment means. The PCV and haemoglobin (Hb) concentration values obtained were within limits for PCV (33.00 – 35.00%) and Hb (9.4 – 17.4 g/dl) reported by Hillyer [19] and Fudge [20], respectively, for growing rabbits. RBC values obtained showed that animals given diet 2 had the highest (5.75) while those fed diet 3 had the lowest (4.50). The values were within the range of 3.07 to 7.50 x 10^6/mm^3 reported by Fudge [20]. The apparent increase in values of PCV, Hb and RBC for rabbits fed diet 2 seems to be an influence of the high dietary protein in yam peel. This corroborates Hackbath [21] and Tuleun et al. [22] who observed that increase in dietary protein could lead to increase in RBC, PCV and Hb. Maxwell et al. [23] reported that blood parameters are important in assessing the quality and suitability of feed ingredients in farm animals. The WBC, MCV, MCH and MCHC values obtained indicated no significant differences (P>0.05) between the treatment groups and that were within limits for WBC (5 to 13 x 10^3 /mm^3), MCV (18 to 24 pg), MCH (50 to 75 um^3) and 26 to 34% for MCHC reported by Anon[24] and Anon (1980). The similarities (P>0.05) for all the haematological parameters in all the treatments and within normal ranges reveal that yam peel, cassava peel and yam – cassava mix meal can successfully replace maize in rabbit diets as far as nutritional and health status of rabbits are concern.

4. CONCLUSION

Based on the comparable results obtained from growth performance and haematological indices, it can be deduced that maize can be successfully replaced with either cassava peel meal, yam peel meal or yam – cassava mix meal. Yam peel meal diet is however preferred since it enhanced weight gain.

COMPETING INTERESTS

Authors have declared that no competing interests exist.

REFERENCES

1. Nwagu FO, Iyeghe – Erakpotobor GT. Energy partitioning for growth by rabbits fed groundnut and stylosanthes forages supplemented with concentrate. Nigerian Journal of Animal Production. 2013;40(1): 37-47.
2. Ekpo JS, Etuk IF, Eyoh GD, Obasi OL. Effects of dietary three sun dried cassava feed forms on the performance of weaner rabbits. Nigeria Journal of Agric. Technology. 2008;13:16-21.
3. Oliver, B. Nutritional value of Rabbits meat. Available:www.buyrabbitmeat.co.za
4. Lebas F, Coudert P, Rouvie R, Rochambeau H. de. The rabbit husbandry, health and production. FAO animal production and health series No. 21, Rome, Italy. 1996;17.
5. Adesina F. (Editor in chief) Daily Sun Newspaper. 2014;10(3047):6.

6. Food and Agricultural Organisation of the United Nations Statistics (FAOSTAT database P. 1D567); 2005.
7. Hahn SK. Cassava research to overcome the constraints to production and use in Nigeria. In proceedings cassava toxicity and thyroid research and public health issues, edited by F. Debage and R. Ahluwalia. Canada; 1983.
8. Tewe OO, Bokanga M. Cost effective cassava plant-based ration for poultry and pigs. In: proceeding of the ISTRC, Africa Branch (IITA). Ibadan, Nigeria; 2001.
9. Ansa OA. Youth participation in agriculture. A paper presented at the agricultural summit organised by ministry of agriculture, Akwa Ibom State, Nigeria on 10th – 12th Feb; 2010.
10. AOAC. Official Methods of Analysis of Official Chemists. 17th Ed, Association of Analytical Chemists, Washington DC, USA; 2002.
11. Monica C. Medical laboratory manual for tropical countries. Microbiology ELBS ed. 1984;11:310-315.
12. Little TM, Hills FJ. Agricultural experimentation – design and analysis. John Wiley and Sons, New York; 1978.
13. Obi IU. Statistical methods of detecting differences between treatment means. Snap press. Enugu, Nigeria; 1990.
14. Xiccato G. Feeding and meat quality in rabbits: A review. World Rabbit Science. 1999;(7):75-86.
15. Summers, J. June 2000. Available:http://www.omafra.gov.on.ca/ene rgyinpoutrydiets.htm
16. Pesti GM. Impact of dietary amino acid and crude protein levels in broiler feeds on biologicalperformance. Journal of Applied Poultry Research. 2009;18:477-486.
17. Berres J, Vieira SL, Dozier WA, Cortes MEM, de Barros R, Nogueira ET, Kutschenko M. Broiler responses to reduced protein diets supplemented with valine, isoleucine, glucine and glutamic acid. J. Appl. Poult. Research. 2010;19:68-79.
18. Olorunsanya B, Ayoola MA, Fayeye TR, Olagunju TA, Olorunsanya EO. Effects of replacing maize with sun-dried cassava waste meal on growth performance and carcass characteristics of meat type

Available:www.sunnewsonline.com   (29th Dec)

rabbit. Livestock Research for Rural Development. 2007;19 (4).

19.   Hillyer EV, Pet Rabbits. The veterinary clinics of North America. Small Animal Practice. 1994;24(1):25-65.

20.   Fudge CS. Laboratory medicine: avian and exotic pets. WB Saunders, Philadelphia, USA; 1999.

21.   Hackbath H, Buron K, Schimansley G. Strain difference in inbred rats: Influence of strain and diet on haematological traits. Lab Anim. 1983;17:7-12.

22.   Tuleun CD, Adenkola AY, Oluremi OIA. Performance characteristic and haematological variables of broiler chickens fed diets containing mucuna seed meal. Trop Vet. 2007;25:74-81.

23.   Maxwell MH, Robertson W, Spencer SCC. Macloroquodale comparison of haematological parameters in restricted and ad libitum fed domestic fowls. British Poult-Sci. 1990;31:407-413.

24.   Anon. Guide to the care and use of experimental animals. Canadian Council of Animal Care. Ontario, Canada. 1980;1:185-190.

10

Comparison of the Nutritional Value of Egg Yolk and Egg Albumin from Domestic Chicken, Guinea Fowl and Hybrid Chicken

L. Bashir[1*], P. C. Ossai[1], O. K. Shittu[1], A. N. Abubakar[1] and T. Caleb[1]

[1]Department of Biochemistry, Federal University of Technology, PMB 65, Minna, Niger State, Nigeria.

Authors' contributions

This work was carried out in collaboration between all authors. Author LB designed the study, carried out the laboratory work and performed the statistical analysis. He equally wrote the first draft of the manuscript and undertook the final editing of the paper. Authors PCO and OKS took part in the laboratory work, part of the draft and undertook in the initial editing of the paper. Authors ANA and TC carried out most of the literature searches and participated in designing the experiment. All authors read and approved the final manuscript.

Editor(s):
(1) Vincenzo Tufarelli, Department of DETO - Section of Veterinary Science and Animal Production, University of Bari "Aldo Moro", Italy.
Reviewers:
(1) Anonymous, Poland.
(2) Anonymous, Slovakia.
(3) Anonymous, Turkey.
(4) Anonymous, USA.

ABSTRACT

The present study was conducted to compare the nutritional and physical quality of egg yolk and egg white of birds from three different genotypes (domestic chicken, hybrid chicken and guinea fowl). The egg yolk and white from each of the bird were separated and analyzed for proximate, vitamins and minerals using standard analytical methods. The eggs of the 3 bird species showed similar conical shape, however, weight of whole egg, egg white and yolk of hybrid chicken was much higher than that of domestic and guinea fowl. The moisture (60.45±0.14%) and vitamin C (121.50±0.14mg/100g) contents of egg yolk were significantly higher in hybrid chicken than in domestic chicken and guinea fowl while the protein (5.47±0.88%), ash (1.32±0.03%) and vitamin C (68.50±0.70mg/100g) contents of egg white was higher in hybrid chicken than domestic chicken and guinea fowl. However, moisture contents (87.45±0.71%) of egg white from guinea fowl was

Corresponding author: E-mail: bashirlawal12@gmail.com;

significantly (p<0.05) higher than hybrid chicken. All elements considered in this study had higher concentrations (mg/100g) in egg yolk than white except for Na whose concentrations were higher in egg white than yolk. The concentration of K^+ (321.50+7.62 and 119.50+2.6.2), Fe^{2+} (12.45+0.09 and 4.45+0.0.8) and Ca^{2+} (26.60+0.63 and 9.23+0.22) for egg yolk and white respectively was significantly (p<0.05) higher in guinea fowl than domestic and hybrid chicken. However, Na contents in hybrid chicken (850.00+22.40 and 975.00+09.00) for egg yolk and white respectively was significantly (p<0.05) higher than that of guinea fowl and domestic chicken. It is concluded that egg yolk and white of hybrid chicken were riches in moisture, protein, ash, vitamin C and sodium than guinea fowl and domestic chicken. While egg yolk and white of guinea fowl were rich in K^+, Fe^{2+} and Ca^{2+} than the eggs of domestic and hybrid chicken.

Keywords: Proximate; minerals; vitamin; egg yolk; egg white.

1. INTRODUCTION

A balanced diet is essential for normal growth, health and preservation of the human body. Eggs have constituted an important part of human diets for centuries because of its high quality protein [1]. They are known to supply the best proteins besides milk [2]. It is also rich in amino-acids, carbohydrates, easily digestible fats and minerals, as well as valuable vitamins [3]. The yolk and white components are all of high biological value and are readily digested [4].

Eggs play important culinary roles and are therefore prepared into different dishes. There are many types of poultry species' eggs consumable as a protein and amino acid supplement [5]. Nigeria has the highest number of poultry farm in Africa. Nigeria presently produces about 300,000 tons of poultry meat per annum officially and 650,000 tons of eggs [6]. A parallel record from Poultry Association of Nigeria (PAN), indicates that Nigeria produces presently above 1.25million tons of egg per year. South Africa is the second producer of eggs in Africa [7]. The question arises whether there are interspecies differences in poultry eggs quality which may affect the nutritive value and quality as human food.

In Nigeria, domestic fowl dominated the poultry industry. Out of 150 million poultry population, 120 million (80%) were indigenous. Domestic fowl constituted 91% of this while guinea fowl, duck, turkey and others were 4%, 3% and 2% respectively [8]. Chicken eggs are the most commonly consumed eggs; they are also an inexpensive single-food source of protein. The yolk and albumin values from literature showed 33.04% (chicken yolk), 32.68% (guinea-fowl yolk); 57.10% (chicken albumin), 50.3% (guinea-fowl albumin) [8].

The guinea fowl is a bird native to the African continent [9]. It derives its name from the Coast of Guinea where it is believed to have originated. The indigenous guinea fowl (Numida meleagris) is widely distributed in Africa where it has distinct popularity among small holder farmers. It is believed that guinea fowls were taken to Europe and America by the Portuguese but in these regions the guinea fowls have been scientifically improved resulting in faster growth rate, bigger body size and enhanced egg laying capacity [10]. Guinea fowl breeding hens produce thicker shelled eggs in comparison to that of a regular chicken [11]. However there is paucity of information on the nutritional qualities of eggs from domestic fowl in comparison with eggs from other poultry species.

In this report, we evaluated the nutritional levels of the proximate, vitamins and minerals composition in egg yolks and egg albumin collected from different bird's species.

2. MATERIALS AND METHODS

2.1 Source of Materials

Freshly-laid egg samples from birds of three (3) different genotypes (domestic chicken, guinea fowl and hybrid chicken) were obtained between 21 and 22 September 2014 from a poultry keeper in Minna, Nigeria. The eggs were analyzed for nutritional compositions between 23 and 26 September 2014. At laying time the domestic fowl were approximately 28 weeks, guinea fowl were approximately 52 weeks and the hybrid chicken were approximately 26 weeks old. The birds' genotype was Hy-Line Brown (hybrid chicken), Pearl (guinea fowl) and normal feathered (domestic chicken). The domestic chicken and guinea fowl had outdoor access all the year round Fed with cereal, hay, clover, vegetables and green crop, according to the season of year. The hybrid chicken ate balanced biocomplete cereal-based mixed fodder with several additives daily.

2.2 Sample Preparation

The egg samples were thoroughly washed with distilled water in the laboratory federal university of technology, Minna. Nigeria. The yolk and albumin were separated by breaking a small part of the egg shell at one end and separating the egg albumin from the yolk.

2.3 Evaluation of Physical Quality of Egg

Egg weight was measured with electronic weighing balance. Subsequently, yolk was separated from the white and weighed separately. The weight of shell was calculated by subtracting the weights of yolk and white from the weight of whole egg.

2.4 Proximate Analysis

Moisture and crude fat content were determined according to the standard methods of A.O.A.C [12]. Ash content was determined at 550°C [13]. Crude nitrogen was determined by Kjeldahl method [13] and crude protein determined by using the formula Crude protein = Crude nitrogen × 6.25 [14]. All the analysis was performed in triplicate.

2.5 Mineral Analysis

The method of A.O.A.C [12] was employed for the determination of mineral content.

2.6 Vitamin Analysis

Vitamin C and Vitamin A composition of each of the sample, were determined by the method of A.O.A.C [12].

2.7 Ethical Clearance

Ethical clearance was given by Federal University of Technology, Minna/Nigeria ethical review board (CUERB) in accordance with international standard on the care and use of experimental animals.

2.8 Statistical Analysis

The data obtained were subjected to Analysis of Variance (ANOVA) using SAS statistical package. Means were separated using Duncan's Multiple Range Test (DMRT). Significance was accepted at P<0.05.

3. RESULTS

3.1 Physical Properties

The eggs of domestic chicken, hybrid chicken and guinea fowl showed similar conical shape with blunt and pointed ends; however, weight of whole egg, egg white and yolk of hybrid chicken was much higher than that of guinea fowl and domestic chicken respectively (Table 1).

3.2 Proximate

The proximate compositions of domestic chicken (*Gallus domesticus*), guinea fowl (*Numida meleagris*) and hybrid chicken are shown in Table 2. The moisture and crude protein content was significantly higher in egg white than yolk while the crude fat and ash content was significantly higher in yolk than in egg white for the three (3) eggs sample. For the egg yolk the moisture and the ash contents was significantly higher in hybrid chicken than in domestic chicken and guinea fowl. The egg yolk from the 3 birds species show no significant difference in there protein and fat content while egg white of the hybrid chicken is significantly lowered in moisture (75.50±0.14%) but higher in ash (1.32±0.03) content as compared with the guinea fowl and domestic chicken. The protein content was significantly (p<0.05) lowered in domestic chicken (3.48±0.91) as compared with guinea fowl (5.81±0.62) and hybrid chicken (5.47±0.88).

3.3 Minerals

Table 3 presents the results of mineral analysis of domestic chicken (*Gallus domesticus*), guinea fowl (*Numida meleagris*) and hybrid chicken. All elements considered in this study had higher concentrations in egg yolk than in the white except for Na whose concentrations were higher in the egg white than in the yolk for all the species considered.

3.4 Vitamins

Table 4 presents the results of vitamins A and C contents of egg yolk and white from domestic chicken (*Gallus domesticus*), Guinea fowl (*Numida meleagris*) and hybrid chicken. Vitamin C content of egg yolk is higher in hybrid chicken (121.50±0.14) and lowest in domestic chicken (97.50±0.71) for the egg albumin the highest concentration of vitamin C was recorded for hybrid chicken (68.50±0.70) and least was recorded for domestic chicken (47.00±2.11).

Table 1. Physical properties of eggs collected from domestic chicken (*Gallus domesticus*), Guinea fowl (*Numida meleagris*) and hybrid chicken

	Weight (g)			
Sample	Whole egg	Egg yolk	Egg white	Shell
Hybrid chicken	72.45±2.41[c]	21.40±1.34[c]	37.13±2.10[a]	13.92±1.01[c]
Guinea fowl	41.34±1.71[b]	13.56±1.08[a]	22.76±1.99[a]	5.02±0.98[b]
Domestic chicken	34.48±1.20[a]	11.87±0.99[a]	20.54±2.90[a]	2.07±0.33[a]

Values follow by the same superscript are not differ significantly at p<0.05, values are Mean ± SEM of triplicate determination

Table 2. Proximate compositions of egg yolks and egg white collected from domestic chicken (*Gallus domesticus*), guinea fowl (*Numida meleagris*) and hybrid chicken

	Proximate (%)			
Sample	Moisture	Protein	Crude fat	Ash
	Egg yolk			
Hybrid chicken	60.45±0.14[b]	3.43±0.88[a]	27.65±0.70[a]	3.42±0.23[b]
Domestic chicken	55.60±0.16[a]	3.85±0.91[a]	30.41±0.22[a]	1.50±0.11[a]
Guinea fowl	50.50.±0.71[a]	4.30±0.62[a]	31.49±0.41[a]	2.15±0.25[ab]
	Egg white			
Hybrid chicken	75.50±0.14[a]	5.47±0.88[b]	2.00±0.03[a]	1.32±0.03[b]
Domestic chicken	86.90±0.16[b]	3.48±0.91[a]	1.09±0.30[a]	0.99±0.21[a]
Guinea fowl	87.45±0.71[b]	5.81±0.62[b]	1.21±0.10[a]	0.70±0.12[a]

Values follow by the same superscript are not differ significantly at p<0.05, values are Mean ± SEM of triplicate determination

Table 3. Minerals composition of egg yolks and egg white collected from domestic chicken, guinea fowl and hybrid chicken

	Minerals (mg/100g)			
Sample	Sodium	Potassium	Iron	Calcium
	Egg yolk			
Hybrid chicken	850.00±22.4[b]	162.00±5.88[a]	3.50±0.70[a]	2.90±0.23[a]
Domestic chicken	150.50±7.16[a]	197.50±3.91[a]	7.10±0.11[b]	14.80±0.91[b]
Guinea fowl	191.00±6.71[a]	321.50±7.62[b]	12.45.±0.09[c]	26.60±0.63[c]
	Egg white			
Hybrid chicken	975.00±09.0[b]	82.00±0.88[ab]	0.90±0.21[a]	0.56.50±0.21[a]
Domestic chicken	172.50±2.16[a]	32.50±0.91[a]	2.10±0.01[ab]	5.80±0.09[ab]
Guinea fowl	199.00±6.71[a]	119.50±2.62[b]	4.45.±0.08[b]	9.23.±0.22[b]

Values follow by the same superscript are not differ significantly at p<0.05, values are Mean ± SEM of triplicate determination

Table 4. Vitamins composition of egg yolks and egg white collected from domestic chicken, guinea fowl and hybrid chicken

Sample	Egg yolks (mg/100g)		Egg yolks (mg/100g)	
	Vitamin C	Vitamin A	Vitamin C	Vitamin A
Hybrid chicken	121.50±0.14[b]	0.33±0.88[a]	68.50±0.70[b]	0.23±0.23[a]
Domestic chicken	110.50±0.16[ab]	0.21±0.91[a]	47.00±2.11[a]	0.11±0.91[a]
Guinea fowl	97.50±0.71[a]	0.31±0.62[a]	52.00±4.09[a]	0.78±0.63[a]

Values follow by the same superscript are not differ significantly at p<0.05, values are Mean ± SEM of triplicate determination

However, no significant differences between the egg yolk and white of the 3 bird's species were found in the content of vitamin A.

4. DISCUSION

4.1 Physical Properties

Generally eggs of birds have oval shape with small differences among species. Despite its small differences, egg shape is considered as an important factor in characterizing bird species. In this study the eggs of domestic chicken, hybrid chicken and guinea fowl showed similar conical shape with blunt and pointed ends, however eggs of domestic chicken is more pointed than the two bird species. Similar findings have been reported for egg shapes of quail and guinea fowl [15]. The significantly higher weight of whole egg, egg white and yolk observed in hybrid chicken as compared to domestic and guinea fowl was obviously due to vast difference in the size of these three bird species. This difference could be attributed to the various feed additives, antibiotics or production stimulants fed to the hybrid chicken but deprived domestic and guinea fowl. The weight of hybrid chicken reported in this study (72.45±2.41) (Table 1) was higher compared to 56.41g reported for naked neck chicken and 40.5g for full feathered chicken [16].

4.2 Proximate Composition

Protein is an essential component of human diet which is needed for the replacement of tissue and supply of energy. Protein deficiency cause growth retardation, muscle wasting, oedema, abnormal swelling of the body and collection of fluid in the body of children [17]. This study revealed low protein contents in three poultry egg species and the little amount presents are more abundant in the egg white, however contrary findings have been reported by [15], who reported more protein contents in egg yolk than white from Quai and guinea fowl. Also in this study no significant difference in the protein content of egg yolk from the three poultry egg species, however the egg white from hybrid and guinea fowl chicken contain more protein than egg white of domestic chicken (Table 2).

Dietary fat functions in the increase of palatability of food by absorbing and retaining flavours. This study also revealed that the eggs yolk and white are good and poor source of lipids respectively. Although no significantly difference in fat contents of egg white from the 3 bird species (Table 2), the low fat contents of egg white is an important consideration for people who suffer from elevated cholesterol level, and can also be recommend as part of weight reducing diets. The lipids contents of egg yolk for the three poultry egg species is high (Table 2), a diet providing 1-2% of its caloric of energy as fat is said to be sufficient to human beings as excess fat consumption is implicated in certain cardiovascular disorders such as cancer and

aging [18]. The fat contents of albumin from domestic chicken and guinea fowl in this study is comparable with that reported for chicken and guinea fowl egg [19].

The ash content gives a measure of total amount of inorganic compounds like minerals present in a food. This study revealed that the egg yolk and white of hybrid chicken contain more ash than domestic chicken and guinea fowl (Table 2). This is an indication that the hybrid chicken will contain more minerals. This finding could be attributed to the variation in feed composition fed the birds. Similarly, low ash content has been previously reported for egg from Quai and guinea fowl [15], and chicken egg (0.91±0.03) [20].

This study revealed that egg white contain more moisture content than the egg yolk, also the egg yolk of hybrid chicken contain more water than those found in guinea fowl and domestic chicken. (Table 2) However high water contents of food have been implicated for low shelf life due to microbial attacked [21].

4.3 Minerals

Calcium helps in the regulation of muscle contraction required by children, infants and fetuses for bones and teeth development [22]. The recommended dietary allowance value of calcium is 600-1400mg/kg [23]. The present study show that both egg yolk and white of guinea fowl contain high amount of calcium as compared to the hybrid and domestic chicken. Considering the importance of calcium, its concentration in egg yolk implies that this can contribute to the amount of dietary calcium. However the level of calcium observed in this study was lower than 38.2mg/100g reported for a whole chicken egg [20]. The recommended daily value for sodium is 1100- 3300mg/kg for adults [22]. Hybrid chicken contains high sodium concentration than the guinea fowl and domestic chicken. However, the concentration of sodium in egg yolk and albumin for all the three bird species observed in this study was lower than 134±20 reported for a whole chicken egg [20].

The enrichment of iron in egg would provide improving the nutrition status of people especially in the risk of iron deficiency or anemia group especially infant, children, pregnant women and socioeconomic groups [24]. The recommended daily requirement of iron for man is 6-40 mg/kg [23]. The egg yolk and white was found to contain iron in concentration within the recommended daily requirement. However, the egg yolk contain more iron than the white, contrary findings have been reported by [25], who reported more iron contents in egg white than egg yolk from snail-eating turtle eggs. Also hybrid chicken was found to have the highest concentration of Iron, this poultry egg species from the result obtained can be used in improving the anaemic condition in iron deficient diabetic patients. Potassium is responsible for nerve action and is very important in the regulation of water, electrolyte and acid – base balance in the blood and tissues [26]. In this study, egg yolk of guinea fowl was found to contain the highest concentration of potassium. The level of potassium in this poultry eggs especially the egg yolksis a good indication that its consumption will enhance the maintenance of the osmotic pressure and acid-base equilibrium of the body [27].

4.4 Vitamins

The Recommended Dietary Allowances (RDAs) for vitamins reflect how much of each vitamin most people should get each day. Results of the present study revealed that the 3 poultry egg species studied contain considerable amount of vitamin A and C (Table 4). However, vitamin C content of egg yolk and white is higher in hybrid chicken compare to domestic and guinea fowl. Vitamins A and C have been reported to have antioxidant properties and may protect body against some forms of cancer [28].

The concentration of vitamins is influenced by genetics, rate of egg production and it varies with the composition of the hen's diet [29]. As the concentration of fat-soluble vitamins in the feed increases, so does the content of vitamins in the egg yolk [30]. However, according to [31] for some vitamins, such as vitamin A, the liver acts as a reservoir so that the concentration in the yolk is buffered against large changes in the diet. This finding is supported by the results of the present study as no significant differences between the egg yolk and white of the 3 bird's species were found in the content of vitamin A (Table 4).

5. CONCLUSIONS

The findings of this study showed that eggs of guinea fowl, domestic and hybrid chicken are rich source of protein, vitamins and appreciable number of some essential minerals. However, from the tested parameters, egg yolk and white

of hybrid chicken were rich in proximate, vitamin C and sodium than eggs of guinea fowl and domestic chicken. While egg yolk and white of guinea fowl were rich in K^+, Fe^{2+} and Ca^{2+} than the other two studied birds.

COMPETING INTERESTS

Authors have declared that no competing interests exist.

REFERENCE

1. Forson A, Ayivor JE. Banini GK, Nuviadenu C, Debrah SK. Evaluation of some elemental variation in raw egg yolk and egg white of domestic chicken guinea fowl and duck eggs. Annals of Biological Research. 2011;2(6):676-680.

2. Vaclavik AV, Christain WE. Essentials of food science springer science business media LLC New York. 2008;205-230.

3. Huopalahti R, López-FR, Anton M, Schade R. Bioactive egg compounds springer-verlag heidelberg. 2007;298.

4. Joel N, Udobi1 CE, Nuria A. Effect of oven drying on the functional and nutritional properties of whole egg and its components. African Journal of Food Science. 2010;4(5):254-257.

5. Trziska T. Processing of eggs in egg science Ed University of agriculture in Wroclaw (Poland). 2000;291-401.

6. USDA. The National Agricultural Statistics Service's Chicken and Eggs' Report typically pages 1 and 9 produced by USDA; 2014.

7. The prospect price and peril of poultry industry in Nigeria – Agriculture. Available:Nairalandwwwnairalandcom/136 0866/prospect-price-peril-poultry-industry

8. Adenowo JA, Awe FA, Adebambo OA, Ikeobi CON. Species variations in chemical composition of local poultry eggs In: Book of Proceedings 26th Annual NSAP Conference 21-25 March 1999 Ilorin Nigeria. 1999;278-280.

9. Smith AJ. Poultry the tropical agriculturist (Revised edition) MacMillan with CTA London UK. 2001;242.

10. Teye GA, Abubakari K. Processing of guinea fowl in the northern region of Ghana the savanna farmer of ACDEP. 2007;8(2):17-20.

11. John CM, Kenanao MM. Effect of egg size on Hatchability of guinea fowl keets. International Journal of Innovative

Research in Science Engineering and Technology. 2013;2:10.

12. Association of official analytical chemist: Official method analytical chemist Washinton DC; 1990.

13. Elinge CM, Muhammad A, Atiku FA, Itodo AU, Peni IJ, Sanni OM. Proximate mineral and antinutrient composition of pumpkin (Cucurbita pepo L) seeds extract Int J Plant Res. 2012;2(5):146-150.

14. Alfawaz MA. Chemical composition and oil characteristics of pumpkin (Cucurbita maxima) seed kernels Resilient Bulietin. 2004;5-18.

15. Dudusola IO. Comparative evaluation of internal and external qualities of eggs from quail and guinea fowl International Research Journal of Plant Science. 2010;1(5):112-115.

16. Rajkumar U, Sharma R, Rajaravindra K, Niranjan M. Effect of genotype and age on egg quality traits in naked neck chicken under tropical climate from India Int J Poult Sci. 2009;8:1151-1155.

17. Mounts TL. The chemistry of components 2nd Edn Royal Society of Chemistry; 2000.

18. Antia BS, Akpan EJ, Okon PA, Umoren IU. Nutritive and antinutritive evaluation of sweet potatoes (Ipomoea batatas) leaves Pakistan Jornal of Nutrition. 2006;5:166-168.

19. Polat ES, Ozcan BC, Mustafa G. Fatty acid composition of yolk of nine poultry species kept in their natural environment Animal Science Papers and Reports. 2013;4:363-368.

20. Matt D, Veromann E, Luik A. Effect of housing systems on biochemical composition of chicken eggs Agronomy Research. 2009;7(Special issue II):662-667.

21. Adeyeye EI, Ayejuyo OO. Chemical composition of Cola accuminata and Grarcina kola seed grown in Nigeria International Journal of Food Science and Nutrition, 1994;45:223-230.

22. Margaret L. Vickery B. Plant Products of Tropical Africa Macmillan in College ed London; 1997.

23. Bolt GH, Bruggenwert MGM. Solid chemistry basic elements Elsevier Scientific publishing Co New York.1978;145.

24. Demmouche A, Lazrag A, Moulessehoul S. Prevalence of anaemia in pregnant women during the last trimester: Consequense for birth weight European Review for Medical and Pharmacological Sciences. 2011;15(4):436-445.

25. Tanasorn T, Wanna T, Wattasit S. Determination of chemical compositions of snail-eating turtle (Malayemys macrocephala) eggs agriculture science developments. 2013;2(4):31-39.

26. National Research Council (NRC). Recommended dietary allowances national academy press Washington DC; 1998.

27. Odoemena CS, Ekanem NG. Antimicrobial assessment of ethanolic extract of Costus afer Leaves. Journal of science and Technology. 2006;5(2):51-54.

28. Wright K. Healing foods geddes and grosset scotland. 2002;8-31.

29. Leeson S, Caston LJ. Vitamin enrichment of eggs J Appl Poult Sci Res. 2003;12:24-26.

30. Sirri F, Barroeta A. Enrichment in Vitamins In: Huopalahti R, López-FR, Anton M, Schade R, (eds): Bioactive egg compounds Springer-Verlag Heidelberg. 2007;21:171.

31. Naber EC. The effect of nutrition on the composition of eggs. Poultry Sci. 1979;58:518-528.

11

Assessing the Use of Indigenous Communication Media among Rural Dwellers of Osun State, Nigeria

S. A. Adesoji[1*] and S. I. Ogunjimi[2]

[1]Department of Agricultural Extension and Rural Development, Obafemi Awolowo University, Ile Ife, Osun State, Nigeria.
[2]Department of Agricultural Economics and Extension, Federal University, Oye-Ekiti, Ekiti State, Nigeria.

Authors' contributions

This work was carried out in collaboration with the authors. The work was designed by the author SAA, carried out the statistical analyses and wrote first draft. Author SIO managed the analyses, corrected the draft by improving on the literature. Both author read and approved the final manuscript.

Editor(s):
(1) Hassen B. Hussien, Department of Agricultural Economics, Wollo University, Ethiopia.
(2) Anonymous.
Reviewers:
(1) Naledzani Rasila, University of Venda, South Africa.
(2) Anonymous, Russia.
(3) Anonymous, South Africa.
(4) Taiye Fadiji, Department of Agricultural Economics and Extension, University of Abuja, P. M. B. 117, Abuja, Nigeria.

ABSTRACT

The study assessed the use of indigenous communication media among rural dwellers of Osun state to determine their knowledge level. Specifically, indigenous communication media that are often utilized by the rural dwellers were identified and socio-economic characteristics were described. Key informants and other 120 respondents were interviewed using interview schedule from two communities in each of the six administrative zones of the state. Data analysis was carried out using frequency counts, percentage, mean, standard deviation. Data analysis showed the mean age of respondents to be 47±7 years and 65% of them were male. Majority were literate. The findings revealed that majority had low knowledge about indigenous communication and more so, most of this communication had gone to extinction except on few cases such as use of proverbs, folklores and songs in which they were highly knowledgeable. Key informant interview showed that coded symbols were phasing out. The study concluded that the use of indigenous communication media was being gradually phased out. It recommends that Indigenous communication media that are often used should not be abandoned in the face of modern communication media, and those that are abandoned should be used so that people, especially the young, would get used to them.

Keywords: Usage; indigenous communication; media; rural dwellers.

1. INTRODUCTION

Indigenous communication media are the vehicles common people or rural dwellers employ for the delivery of their messages. [1] viewed traditional media as body languages and other non-verbal languages being used in the traditional societies for a variety of purposes. Indigenous communication systems are means by which local people communicated with one another in the primitive era. These systems of communication are passed from one generation to another. Such communication systems are derived from society's experience and thoughts over a long period of years. The messages and ideas are transmitted by means of itinerant dance and mime groups, puppet shows and other folk media which serve not only to entertain but to influence attitudes and behaviour' [2]. Before the introduction of printed material, radio, film and television, mass communication in Nigeria was done through the indigenous systems of communication. [3] claimed that Western commerce, religion, education, politics and the form of government, imported from Britain from the 19th century, found the indigenous communications systems inadequate for several reasons: the systems use local languages, they are interactive in the form of several chains of face to-face activities from the source or sources to the receivers, they are exclusively integrated into local cultures; they do not depend on Western technology and they are dissimilar from the Western model of mass communication [3]. Despite the deficiency in indigenous communication, this type of communication system has been able to sustain the rural dwellers before the introduction of modern means of communication. It provides identity among communities that uses them and also guides unruly behaviours in the society. It must be noted that the expression indigenous communication in this context is not a substitute for archaic, barbaric or rudimentary communication systems; rather it represents traditional and ancient communication systems [4]. It is no exaggeration that the average African man today is, to some extent, a deculturalised person, living on foreign cultural values. Though no one should be blamed for this development, It

is brought about by several factors such as colonization, Introduction of foreign religions - Christianity and Islam, social change, travelling, urbanization etc which brought about diffusion of cultures. Arising from this situation, many indigenous communication systems were condemned by westerners and were looked upon with disdainful eyes to the extent that local people became discouraged in using them as they were seen as archaic, traditional, rudimentary and sometimes fetish. [5] reported that westerners believed most of the indigenous communication media have a religious background and were thus condemned with traditional religion.

There is often a certain degree of semantic and conceptual confusion and misapprehension surrounding what constitutes indigenous or traditional communication, It was stated by [6] that indigenous communication as used in Africa is an admixture of social conventions and practice that have become sharpened and blended into veritable communication modes and systems which have almost become standard practices for society. It can be said to be a complex system of communication, which pervades all aspects of rural life. Indigenous communication was a pre-colonial medium/channel of communication which has its own limitations especially in the area of national commerce, education, politics and government [6]. Indigenous communication is perhaps the most important way by which the rural dwellers communicate among themselves and with others. According to [7,8] interest in the use of indigenous media is now increasing in less developed countries as a credible and acceptable source of information because mass media have been less successful in promoting rural development.

Indigenous communication media such as theatre, drama, puppet shows, drumming, village criers, storytellers, orators, songs, using a bell, folk tales, proverbs, ceremonial occasions like initiations, funerals, wedding, announcements etc have played and continue to play important roles among rural and poor communities [9,10,11]. It is important to note that these indigenous communication systems are peculiar to specific societies; borne out of the people's culture, religious conviction and experiences and so their interpretations may vary from one society to the other. In any case, they do reveal the ethics of each society. [12] defined indigenous media as "any form of endogenous communication system which by virtue of its origin from/and integration into a specific culture, serves as a channel for messages in a way and manner that requires the utilization of the values, symbols, institutions and ethos of the host culture through its unique qualities and attributes" [5] opined that indigenous media should be encouraged because it gives room for local participation in development efforts and since exogenous media fail to reach the rural dwellers.

For generations, rural populations living in isolated villages without access to modern means of communication have relied on the spoken word and indigenous forms of communication as a means of sharing knowledge and information and providing entertainment. In some rural places where modern communication systems have not reached, the dwellers make use of the indigenous communication systems and pass same to their offspring through generations by socialisation processes. Coded information, drums, folklores are still being used both in urban and rural areas. These are common among the aged and traditional members of the society. However, these media of communication are phasing out among the youths and elites of the society. In the light of the above, this study seeks to provide answers to the following research questions.

(i). Are the indigenous communication media in use again in the rural communities of Osun State?
(ii) What is the knowledge level of rural dwellers in the use of indigenous communication media?
(iii) Which of the popular communication media are still been used in the rural communities of the state?

1.1 Objectives of the Study

The main objective of the study is to assess the use of indigenous communication media among rural dwellers of Osun State.

Specific objectives include:

(i) Describe socio-economic characteristics of the rural dwellers in the study area
(ii) Determine the perceived knowledge level of rural dwellers in the use of indigenous communication media in the State.
(iii) Identify the indigenous communication media that are often utilized by the rural dwellers in the State.

The study also hypothesized that there is no significant relationship between the Use of Indigenous Communication Media and the socio-economic characteristics of Rural Dwellers and that there is no significant relationship between the Use of Indigenous Communication Media and knowledge level of rural dwellers.

2. METHODOLOGY

The study was conducted in the six Administrative zones of Osun State. It is located in the south-western region of the country. The State is believed to be rich in culture and tradition, and the major occupation of the people of the State is farming. The state lies within the rainforest region on Western Nigeria between latitude $6^\circ 50^1$ and $8^\circ 10^1$ on the north-south pole and longitude $4^\circ 05^1$ and $5^\circ 02^1$ on the east-west pole and is within the tropics bounded by Kwara State in the north, Ogun State in the south, Oyo State in the west and in the east by Ondo State. The population of the study areas consists of rural dwellers either involved in farming or non-farming activities.

Multistage sampling procedure was used to select respondents from the four states. In first stage, all the six administrative zones, namely; Ife, Ilesa, Ikirun, Osogbo, Ila, and Iwo, were selected for the study. Two rural communities were selected from each zone making a total of twelve communities. Ten households were randomly sampled from each of the community and the household heads were interviewed for the study. A total of one hundred and twenty respondents were interviewed using interview schedule. In addition, two key informants were interviewed in the state.

Two variables were investigated in the study, dependent and the independent. The dependent variable is the use of indigenous communication media. This was measured using regularly used and not used. Also the knowledge level was measured with four point scale of high knowledge, moderate knowledge, low knowledge and no knowledge. These were appropriately scored. Mean ± standard deviation was used to categorize knowledge level into high, medium and low. The independent variables in the study include sex, level of education, marital status, religion and occupation of the respondents. Descriptive statistics such as frequency, mean, percentage, standard deviation were used to summarize the data.

3. RESULTS AND DISCUSSION

Results in Table 1 show that majority (65.8%) of the respondents were male. About 12.5% were below the age of 26 years with 31.7% between 26 and 45 years while 45.8% were between 46 and 65 years and 10% were above 65 years. Inspite of the fact that most of the rural dwellers are still in their prime age with old people also dwelling in the study area it is expected that the indigenous knowledge will be encouraged but reversed is the case. The mean age of the respondents was 47 years with standard deviation of 7. The implication of this is that rural dwellers in the study area were still in their productive ages. The results in the table further show that about 72% of the respondents were married, 13.8% were single with 11.7% widowed. This result support the report of [12,13,14] that majority of responsible adults in rural areas were married. The mean household size of the respondents was 7.1 with standard deviation of 2.4. A little above average (55.8%) had household size of between 6 and 8 members while 28.3% had household size between 1 and 5 members.

All members of the household belong to one religious group or the other. Only 4.2% practiced traditional religion, 42.5% practiced Christianity while 51.6% were practicing Islamic religion. The implication of this is that majority practiced foreign religion which might have influence on the use of indigenous communication in negative ways because of culture of assimilation. This is in-line with [5] assertion that westerners believed most of the indigenous communication media have a religious background and were thus condemned with traditional religion. Above average (53.3%) of the respondents were indigenous members of their respective communities while 44.2% were non-members. This shows that rural communities were accommodating and that they allowed strangers in their mist. The mean years of residence of the respondents in their various communities were 24.3 years with a standard deviation of 7.2. About 26.7% had lived between 11 and 20 years

while 18.3% lived between 1 and 10 years and 4.2% lived above 50 years in their various communities. The educational status of the respondents showed that 18.3% of the respondents had no formal education, 22.5% had only primary education, and 35.0% had secondary education while 22.5% had education at the tertiary level. The educational status of the respondents might influence their perception towards indigenous communication because western education might have changed their orientation about indigenous communication.

3.1 Perceived Knowledge Level of Respondents in the Use of Indigenous Communication Media

Results in Table 2 show that 24.8% of the respondents had high knowledge level, 57.3% had moderate level of knowledge and 17.9% had low knowledge level in the use of Song as an indigenous communication medium. Results in the table also show that 17.1% of the respondents had high knowledge level in the use of various indigenous communication media. Dances, 36.8% of the respondents had moderate knowledge level, and 45.3% of the respondents had low knowledge while 0.9% of the respondent had no knowledge in the use of Dances as a communication medium. The results show that 55.6% of respondents had high knowledge level in the use of Parable/proverb as a medium of communication while 40.2% had moderate knowledge and 4.3% had low knowledge level. The table further shows that 38.8% of respondents had high knowledge level in the use of Folklores as a medium of communication while 37.1% had moderate knowledge level and 21.6% had low knowledge level in the use of Folklores. Only 2.2% of the respondents had high level of knowledge in the use of Masquerade as a communication medium while 5.7% had moderate level of knowledge and 42.2% had low knowledge level while 50.0% had no knowledge at all in the use of Masquerade.

Results in Table 2 show that 24.8% of the respondents had high knowledge level in the use of Festivals as a communication medium, 36.8% had moderate and low knowledge levels, respectively and 1.7% had no knowledge at all in the use of Festivals. The table also shows that 7.7% of the respondents had high knowledge level in the use of Drama as a communication medium, 15.4% had moderate knowledge level, and 70.9% had low knowledge level while 6.0%

had no knowledge in the use of Drama. The table further show that 19.0% of the respondents had high knowledge level in the use of Poetry as a communication medium while 13.8% had moderate knowledge level and 49.1% had low level while 18.1% had no knowledge in the use of Poetry. Only 17.2% of respondents had high knowledge level in the use of Coded message as a communication medium while 31.0% had moderate knowledge level and 34.5% had low knowledge and 17.2% had no knowledge in the use of Coded messages. Knowledge level in the use of Symbol as a communication medium was high among 9.8% of the respondents, 18.8% had moderate knowledge level while 34.8% had low knowledge level and 36.6% had no knowledge.

Table 1. Distribution of respondents by socio-economic characteristics n=120

Variable	Frequency	Percentage (%)	Central tendency (mean)	
Sex				
Male	79	65.8		
Female	41	34.2		
Age				
0-25 years	15	12.5		
26-35 years	23	19.2	X	47.0
36-45 years	15	12.5	SD	7.0
46-55 years	25	20.8		
56-65 years	30	25.0		
66 years and above	12	10.0		
Marital status				
Single	19	13.8		
Married	86	71.7		
Widowed	14	11.7		
Divorced	1	0.8		
Household size				
1-5 individuals	34	28.3		
6-10 individuals	67	55.8	X	7.1
11-15 individuals	10	8.3		
> 16 individuals	1	0.8		
Missing	8	6.7		
Religion				
Christian	51	42.5		
Islam	62	51.6		
Traditional	5	4.2		
Missing	2	1.7		
Indigenous status				
Yes	64	53.3		
No	53	44.2		
Years of residence				
0-10 years	22	18.3		
11-20 years	32	26.7	X	24.3
21-30 years	23	14.2	SD	7.2
31-40 years	16	13.3		
41-50 years	16	13.3		
50 years and above	5	4.2		
Missing	6	5.0		
Educational status				
Primary	27	22.5		
Secondary	42	35.0		
Tertiary	27	22.5		
No formal education	22	18.3		
Missing	2	1.7		

3.2 Use of Some Indigenous Communication Media

Results in Table 3 show that respondents still make use of the indigenous communication media. For example more than 90% of the respondents make use of these media regularly; Songs (98.3%), Proverb / Parables (99.1%), Festival (97.4%), Folklores (90.5%), also more than 80% of the respondents used the following media regularly, Dance (86.8%), Poetry (80.2%). About 63% regularly use drama as a medium of indigenous communication and 49.6% of the respondents regularly used town crier. Mass media are regularly used to communicate information which drama and town criers could have been used in rural communities. This may account for the low use of indigenous communication media.

Furthermore, those indigenous communication media that are not regularly used include the use of Masquerade (91.5%), Coded message (87.2%), Movement of sun/shadow (98.3%), Sound of insects/Birds (87.7%) and Symbols (72.6%). The findings revealed that majority had low knowledge about indigenous communication and more so, most of this communication had gone to extinction. That view was in line with key informant report from various rural areas.

3.3 Report of the Key Informant

Aworo of Ila – Orangun was interviewed on the use of indigenous media in his area. He had this to say:

The Yoruba culture is fast being eroded by foreign culture. The use of coded message is commonly used by elders and the herbalists. It is also common among members of particular groups and cults. It was regularly used many years back especially among the members of rural communities. In case of masquerades, there are some that are used to carry special sacrifice. They are not usually used during the day light. Symbols are still used by old or elderly people in rural areas. This is to pass information to people with fore knowledge about ones movement. Many homes today do not even speak our language, how do you expect they will know our culture? The way out is to include the teaching of our culture in schools.

Olowaa of Owaa was also interviewed on the use of indigenous communication media. He had this to say:

Indigenous communication media was the only medium of communication before the whites brought their own. But we have left our own and embraced theirs. There are reasons for this. However, our tradition should not die. Even some of the indigenous media that are used in the palace like trumpets to remind the king of events, the traditional drums to remind the king of an outing are not used again, except on special occasions. The use of some of these is left with the aged which may pass out with them. For example, some old people still use shadow to know the time of the day, some still use symbols on the way to their farms. In the case of coded messages, it is still found among traditional title holder, Obas, some chiefs as well as the herbalists (Babalawo) and some traditional worshipers. The way out is to allow our children to know our traditions and practice our culture. It should be made compulsory in school curriculums.

Aworo is the head of herbalists in Igbomina land. His report showed that the use of some indigenous media is limited to the aged and the herbalist.

Report from another key informant, Olowaa of Owaa is the King and traditional head of Owaa community. He also affirmed that many of the indigenous media are left with people that practice the native cultures and adhere to the traditions. It is also to be noted that erosion of the indigenous media is also touching the source of tradition, which is the palace.

3.4 Hypotheses of the Study

Table 4 shows that there were negative and significant relationship at $P \leq 0.5$ between use of indigenous communication media and some socio-economic characteristics of the respondents such as educational status (r=-0.52) and age (r=-0.33). It indicates that younger the age of rural dwellers the less the usage of indigenous communication and vice visa. Moreover, the higher the level of education status of rural dwellers, the less the usage of indigenous communication. The implication of the finding is that rural dwellers with high educational level and younger age were not conversant with the use of indigenous communication. However, significant relationship exists between source of information (r=0.61) and use of indigenous communication. This indicted that availability of reliable source of information about indigenous communication

Table 2. Showing the perceived knowledge level of the respondent in the use of some indigenous communication media

Variable	High knowledge		Moderate knowledge		Low knowledge		No knowledge at all	
	Freq.	%	Freq.	%	Freq.	%	Freq.	%
Songs	29	24.8	67	57.3	21	17.9	0	0.0
Dance	20	17.1	43	36.8	53	45.3	1	0.9
Proverb/parable	65	55.6	47	40.2	5	4.3	0	0.0
Folklore	45	38.8	43	37.1	25	21.6	3	2.6
Masquerade	5	2.2	13	5.7	97	42.2	115	50.0
Festival	29	24.8	43	36.8	43	36.8	2	1.7
Drama	9	7.7	18	15.4	83	70.9	7	6.0
Poetry	22	19.0	16	13.8	57	49.1	21	18.1
Coded message	20	17.2	36	31.0	40	34.5	20	17.2
Symbols	11	9.8	21	18.8	39	34.8	41	36.6
Movement of sun	21	18.8	20	17.2	16	13.7	60	51.3
Sound of insects/birds	6	5.1	8	6.8	32	27.4	71	60.7
Town criers	84	72.4	20	17.2	10	8.6	2	1.7

Source: Field survey, 2011

Table 3. Showing the use of some indigenous communication media

Variable	Regularly used		Not regularly used	
	Freq.	%	Freq.	%
Coded message	15	12.8	102	87.2
Songs	115	98.3	2	1.7
Proverb/parable	114	99.1	1	0.9
Movement of sun	2	1.7	113	98.3
Sound of insect/bird	14	12.3	100	87.7
Festival	114	97.4	3	2.6
Dance	99	86.8	15	13.2
Town crier	56	49.6	57	50.4
Masquerade	10	8.5	107	91.5
Folklore	105	90.5	11	9.5
Poetry	92	80.2	23	19.8
Drama	73	62.9	43	37.1
Symbols	32	27.4	85	72.6

Source: Field survey, 2011

Table 4. Results of correlation analysis showing relationship between use of indigenous communication and some socio- economic characteristics and knowledge level of rural dwellers

Variables	Correlation (r)
Source of information	0.61**
Educational status	-0.52*
Age	-0.33*
Knowledge level of rural dwellers	0.42**

*Source: Field survey, 2011; ** Positive Significant at p≤0.05 * Negative significant*

usually had positive influence on level of usage. Furthermore, there is a significant relationship between the usage of indigenous communication and rural dwellers knowledge level (r=0.42). Implication of the finding is that when rural dwellers were knowledgeable about indigenous communication they tend to use it often.

Correlation analysis between socio-economic characteristics, knowledge level and use of indigenous knowledge.

4. CONCLUSION AND RECOMMENDATIONS

Indigenous communication is part of culture. It is being practiced in places where the intensity of use of exogenous media is low. It is also common among the aged, traditional worshippers and people that value cultures. However, it is gradually phasing out, due to diffusion of exogenous media.

4.1 Recommendations

The traditional rulers and elderly people should endeavour to preserve the culture and tradition of their communities.

Families should introduce their young ones to the use of indigenous communication by using such in their presence and encouraging them to use them.

Indigenous communication media that are often used should not be abandoned in the face of modern communication media and those that are not used should be used so that people, especially the young ones would get used to them.

In addition, indigenous communication media should be introduced in the school curriculum so as to encourage the young ones to see value in the use of indigenous communication.

COMPETING INTERESTS

Authors have declared that no competing interests exist.

REFERENCES

1. Sulaiman A. Osho Oramedia – African means of Communication in a Contemporary World Seminar on Cultural Diplomacy In Africa (CDA), and International Conference on Cultural Diplomacy In Africa – Strategies to Confront The Challenges of the 21st Century: Does Africa Have What is Required? Institute for Cultural Diplomacy (ICD) in Berlin, Germany; 2011.

2. Van der Stichele P. Folk and traditional media for rural development: A workshop held in Malawi, SD: Knowledge: Communication for development; 2000.

3. Segun Oduko. From indigenous communication to modern television: A Reflection of political Developmentin Nigeria Africa Media Review. 1987;1(3).

4. Emerean MB. Style and Meaning in an Oral Literature. Language. 1966;42(2):328.

5. Dzurgba A. Towards the pedagogy of oral ethics in Africa in context of social and cultural studies. 1999;3(1).

6. Olulade R. Culture and communication in Nigeria interpreter magazine, Lagos: Nigeria. 1988;2.

7. Van dan Ban A, Hawkins W. Agricultural Extension, 2nd Edition Longman scientific and Technical Essex.cm 202 IE England; 1996.

8. Yahaya MK. Development Communication: Lessons from change and Social Engineering projects, Ibadan Corporate Graphics Limited; 2003.

9. Kamlongera CF, Mwanza WB. An anthology of Malawian literature for Junior Secondary, Dzuka Publishing Company, Blantyre; 1993.

10. Mundy P, Compton JL. Indigenous communication and indigenous knowledge. Development Communication Report. 1991;74.

11. Mushengyezi A. Rethinking indigenous media: Rituals, talking drums and orality as forms of public communication in Uganda. Journal of African Cultural Studies, Clearinghouse on Development Communication, Arlington, VA. 2003;16(1): 107-117.

12. Ansu-Kyeremeh K. Perspectives in indigenous communication in Africa: Theory and application, school of communication studies, University of Ghana, Legon; 2000.

13. Jibowo OO. Essential of rural sociology. Gbemi Sodipo Press Ltd. Abeokuta, Nigeria. 1992;54.

14. Ipaye GA. Analysis of role performance of contact farmers in T & V. extension system of Lagos State ADP (Unpublished Ph. D). thesis. Department of Agricultural Extension Services, University of Ibadan, Nigeria. 1995;8.

12

Effect of Water Volume on Growth, Survival Rate and Condition Factor of *Clarias gariepinus* Hatchlings Cultured in Different Enclosures

S. A. Okunsebor[1*], K. G. Shima[1] and T. F. Sunnuvu[2]

[1]*Department of Agriculture, Shabu-Lafia Campus, Nasarawa State University, Keffi, P.M.B. 135, Lafia, Nasarawa State, Nigeria.*
[2]*School of Agriculture, Lagos State Polytechniques, Ikorodu Lagos, Nigeria.*

Authors' contributions

This work was carried out in collaboration between all authors. Author SAO designed the study, wrote the protocol, and wrote the first draft of the manuscript and laboratory work. Authors KGS and TFS performed the statistical analysis, managed the analyses of the study and literature searches. All authors read and approved the final manuscript.

Editor(s):
(1) Yeamin Hossain, Department of Fisheries, Faculty of Agriculture, University of Rajshahi, Bangladesh.
Reviewers:
(1) Anonymous, Czech Republic.
(2) Rakesh Kumar Pandey, Department of Zoology, Kamla Nehru Institute, Sultanpur 228118, India
(3) Anonymous, Nigeria.
(4) Anonymous, Egypt.

ABSTRACT

Fry management in aerated, none aerated aquarium and hapa system were determined in Fish Hatchery of Faculty of Agriculture, Shabu Lafia, Nasarawa State University Keffi, Nigeria to assess condition factor, percentage survival rate, increase in total body length and percentage weight gain. Two hundred hatchlings each of *Clarias gariepinus* were put into12.6 litres of water of 35 x 30 x 15 cm aquarium (with aerator and without aerator) and those of hapas (35 x 30 x 15 cm dimensions) were placed each in 1000 litres of water (aerated and none aerated) in 3 replicates. The fry were fed at 5% of their body weight with Artemia shell free as fry conventional food. The feeding was done four times daily at ¼ part of the 5% body weight for the period of sixteen days. Temperature (27.45±0.05°C), pH (7.56±0.03); dissolved oxygen (8.20±0.03 mg/L), total alkalinity (15.36±0.03 mg/L) and free carbon dioxide (4.30±0.03 mg/L) monitored in the various treatments were not significantly different from each other. The Percentage weight gain (1117 and 1067), percentage survival rate (92.83 and 91.33), increase in total body length (1.07 and 1.07cm) and condition factor (11.99 and 11.44) of *the* fry in hapa system (aerated and none aerated respectively) were significantly ($p < 0.05$) higher than those of aquaria treatments. The results of aerated and none aerated hapa treatments were not significantly different ($P > 0.05$) from each other. The use of hapa as improved system for mass production of *C. gariepinus* fry is highly recommendable in the large body of water.

Keywords: Aquarium; aerator; fry; hatchlings; hapa; Clarias gariepinus.

1. INTRODUCTION

Clarias gariepinus is one of the widely cultivable fish in Nigeria due to acceptability and its resistance to poor water quality [1]. In spite of remarkable success on the hatching of *Clarias gariepinus,* the survival at fry stage is still a limiting factor [2]. This can be attributed to lack of proper awareness and technicality involved in principles of hatchery management especially the problem of ammonia accumulation and fungi from waste food and faecal materials. The fry of *Clarias gariepinus* fish is mass produced in most hatcheries but the percentage survival rate of the fry is low compared to the number of hatchings in every hatchery operation [3]. Identification of some causes for the low and variable survival of *C. gariepinus* fry was reported in Cameroon to be predation (primarily by amphibians and aquatic arthropods) and cannibalism (exacerbated by low food availability) [3]. The mortality rate of *C. gariepinus* fry is still significantly high even when factors like predation and cannibalism are controlled. Therefore, there is need for solution to the problem of fry mortality in the hatchery especially in the first two weeks of active live. The uses of Hapa for mass production of *C. gariepinus* fry need to be investigated especially when predation and cannibalism factors are controlled [2]. Nylon or "mosquito" net cages commonly referred to as hapas or net-hapa hatchery/nursery system has been reported to be very efficient for the production of high quality *Tilapia* fish seeds for stocking ponds [4]. Performances of Dutch *Clarias* juvenile stocked at different densities in out-door hapas was also reported in feeding experiment by [5]. The use of hapa for *Clarias gariepinus* feeding experiment was also documented by [6]. In this study, the efficiency of hapa system in high volume of water in mass production of *C. gariepinus* fry is examined to assess its effects on condition factor, percentage survival rate, increase in total body length and percentage weight gain of the fish in sixteen days.

2. MATERIALS AND METHODS

Two hundred hatchlings of *C.gariepinus* fry were put into 12.6 litres of water in 35 x 30 x 15 cm glass aquarium with aerator (Awa), glass aquarium without aerator (awta) while hapa with aerator (hapwa) and hapa without aerator (hapwta) (size 35 x 30 x 15 cm each) were placed in 1000 litres of water. The whole set up of the treatments were in 3 replicates. In the construction of hapas, materials and methods reported by [2] was employed. The hapas were made of brown colour plankton net and they were submerged to same depth of water level like those of the aquaria. The fry were fed with Artemia shell free four times daily at ¼ part of 5% of their body weight (to avoid cannibalism and to increase food availability) for the period of sixteen days. The condition factor, percentage weight gain, increase in total body length (cm), and percentage survival rate were determined as follows:

i.   Condition factor (k) = (weight gain (g) / L^3) x 100 Where L = total length in cm attained during the experiment [7].
ii. Increase in total body length in cm = final length – initial length at the start of the experiment [8].
iii. Survival rate (%) = (No. alive after the experiment/Total No. of fry at the start of the experiment) x 100
iv. Percentage weight gain = (final weight (g) – initial weight (g) /initial weight (g)) x 100 [9].

The weight was taken using digital sensitive scale while the increase in total length was taken using a tape rule graduated in millimeter and centimeters. Cannibalism was eradicated by feeding the fry adequately and timely. All dead fry were collected and recorded on daily bases just before feeding using siphoning method [10]. Replacement of ⅔ of the water in aquarium was done daily. Temperature (°C), pH, alkalinity (mgL^{-1}), free Carbon oxide (mg/L) and dissolved oxygen (mL^{-1}) in water used for each treatment were collected in depth of 2.00 cm below the water surface. The use of thermometer, pH meter and the methods of [11] were employed for the determination of water quality parameters in this experiment. The data obtained were analyzed using descriptive statistics, analysis of variance and Duncan's multiple range tests for the level of significant difference between means at probability of 0.05.

3. RESULTS AND DISCUSSION

3.1 Water Quality Parameters

Table 1 shows water quality parameters of source of water for all treatments monitored in this study. The temperature for each of the treatment 'Awa', 'Awta', 'Hapwa' and 'Hapwta' in all the treatments were not significantly different (p>0.05) from each other. Similar results of insignificant difference were recorded for Carbon dioxide, Total alkalinity, Dissolved oxygen and pH throughout the period of the experiment. Results of each treatment were not altered by water parameters in this experiment as they were not significantly different in the treatments and the water was from same source. The average temperature, dissolved oxygen, total alkalinity and carbon dioxide for the various treatments observed in this study were within acceptable range [12].

3.2 Percentage Weight Gain

Results in Fig. 1 shows that the percentage weight gain of the C. gariepinus fry were not significantly different in the aerated and none aerated hapa. The aerated and none aerated hapa treatments significantly (P<0.05) had the best weight gain (%) in this experiment. The aerated and none aerated aquaria were significantly the same in percentage weight gain. Regular changing of (⅔) of the volume of water in the aquarium could have minimize deterioration of the water quality in the in the aquarium but in mass production of fry, hapa system will lessens effort and inputs with better results compared to aquaria tanks.

3.3 Percentage Survival Rate

The results of percentage survival rate of this experiment are shown in Fig. 2. The Hapa system (Hapwta and Hapwa) and aquaria (Awa, Awta) were significantly different from each other (P<0.05). Although the survival rate of fry was

very encouraging and results were very close for the Hapwta and Hapwa (92.83 and 91.33 respectively), the results of Hapwa was not significantly better than Hapwta in large volume of water. In a small volume of water, such as in small tanks there is need of aeration as the dissolved oxygen can be depleted by the fry. In all the results, hapa system, whether aerated or not, stands significantly the best in the production of C. gariepinus fry because it takes care of ammonia level, predators, and aquatic arthropods [2] which may directly and indirectly affect the fry. The least value of survival rate (%) of fry was in aquarium without aerator because of dissolved oxygen utilization. This is clear indication that additional accessories are needed to run the none hapa system in successful rearing of C. gariepinus fry.

3.4 Increase in Total Body Length (cm)

The results of the fry increase in total body length during the period of the experiment are shown in Fig. 3. Aerated and none aerated hapa were significantly same (P>0.05) but they were significantly (P<0.05) better than the results of increase in the body length of those of aerated aquarium tanks. The large water volume might be the major factor helping the water quality for growth as others factors like predation [3] and cannibalism were controlled. This investigation also indicated that growth in length can be achieved in hapa system.

3.5 Condition Factor

Results of the effects of treatments on the condition factor of the fry are shown in Fig. 4. C. gariepinus fry in Hapa with aerator and Hapa without aerator significantly had the highest condition factor (11.99,11.44). They were significantly different (P<0.05) from other treatments of the experiment. The least condition factor (8.18) was found in aquarium without aerator and it is not significantly different from the aquaria with aerator. The regular changing of ⅔ of the water daily in the aquarium must have minimized the adverse effect of the none aeration of the aquarium, hence the none significant difference recorded. The Hapa system produced a good result because of the large volume of water and the plankton net that allow for the ammonia from waste food and faecal material to go out of the hapa. Holding fry in hapas to protect them from both amphibians and aquatic arthropods, was reported to have decreased mortality by 5.7 percent and installation of bird-netting over the hapas reduced mortality by 21.7 percent [3]. Although the fry in hapa were restricted to the same size of 35 x 30 x 15 cm like those in aquaria, the water volume (1000 litres of water) from the surrounding normalizes the water quality in the Hapa. The hapa with aerator yielding significantly

(p>0.05) the same result with hapa without aerator, indicates that in hapa system of large body of water, dissolved oxygen in the water can naturally sustains the given fry. This results show that with or without aeration, the fry can survive in hapa system in a large volume of water.

Table 1. Water quality parameters of source of water for treatments in this study.

Parameter	Awa	Awta	Hapwa	Hapwta
Temperature (°C)	27.43±0.04	27.43±0.06	27.44±0.02	27.43±0.05
pH	7.45±0.02	7.46±0.03	7.47±0.03	7.46±0.03
Total alkalinity (mg^{-1})	15.21±0.03	15.21±0.01	15.21±0.02	15.21±0.03
Dissolved oxygen (mg^{-1})	8.20±0.03	8.21±0.03	8.20±0.03	8.20±0.01
Carbon dioxide (mg^{-1})	4.20±0.01	4.20±0.03	4.20±0.02	4.20±0.03

Awa = Aquarium with aerator, Awta = Aquarium without aerator, Hapwa = Hapa with aerator, Hapwta = Hapa without aerator

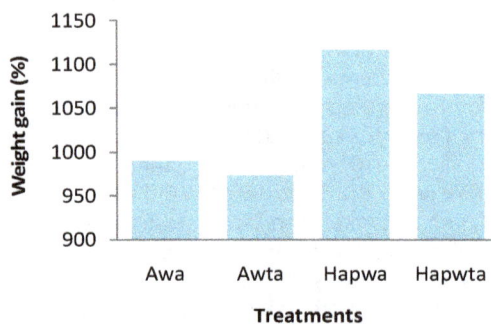

Figure 1: Effect of treatments on weight gain (%) *Clarias gariepinus* fry

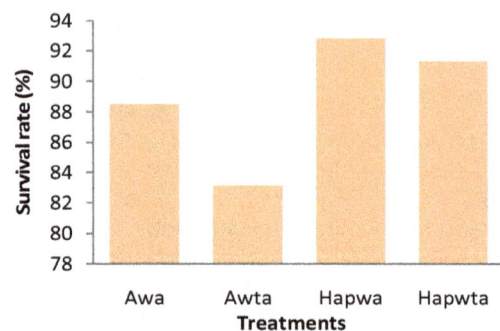

Figure 2: Effect of treatments on survival rate (%) *Clarias gariepinus* fry

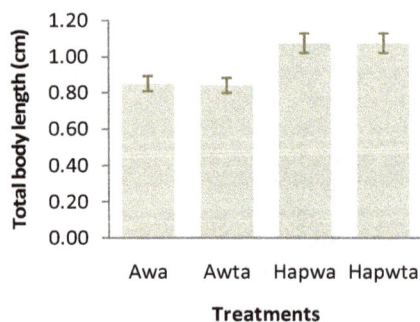

Figure 3: Effect of treatments on increase in total body length (cm) *Clarias gariepinus* fry

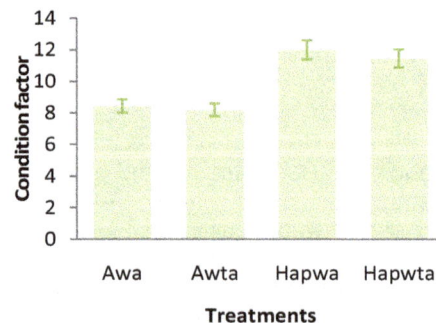

Figure 4: Effect of treatments on condition factor of *Clarias gariepinus* fry

Awa = Aquarium with aerator, Awta = Aquarium without aerator, Hapwa = Hapa with aerator, Hapwta = Hapa without aerator.

4. CONCLUSION AND RECOMMENDATIONS

The use of hapa in mass production of *C. gariepinus* fry is significantly better than the aquarium especially in large volume of water body. The Percentage weight gain, percentage survival rate, increase in total body length and condition factor of *C. gariepinus* fry in hapa system is significantly higher than those studied in aquaria. The use of hapa as an improved system for mass production of *C. gariepinus* fry is highly recommendable in the large body of water.

ACKNOWLEDGMENTS

The authors appreciate Department of Forestry, Wildlife and Fisheries, Faculty of Agriculture, Nasarawa State University Keffi, for providing hatchery for this investigation.

COMPETING INTERESTS

Authors have declared that no competing interests exist.

REFERENCES

1. Okunsebor SA, Sotolu AO. Growth performance and survival rate of *Clarias gariepinus* fry fed on live feeds *Brachionus calyciflorus, Ceriodaphnia reticulata* and shell free Artemia. PAT. 2011;7(2):108-115

2. Nguenga D, Forbin I, Teugles GG, Ollevier F. Predation capacity of tadpoles (*Bufo regularis*) using African catfish *Heterobranchus longifilis* larvae: Impact of prey characteristics on vulnerability to predation. Aquaculture Research. 2000;31:931-936.

3. Sulem SY, Brummett RE. Relative importance of various predators in *Clarias gariepinus* fry mortality in Cameroon. NAGA, World Fish Center Quarterly. 2006;29(3&4):74-77

4. Otubusin SO, Opeloye G. Studies on fish production using floating bamboo net-cages and hapas in Kainji Lake Basin. Kainji Lake Research Institute Annual Report. 1985;82–87.

5. Sogbesan OA, Aderolu ZA, Panya MW. Performances of dutch claries juvenile stocked at different densities in out-door happas. World Rural Observations. 2009;1(1):17-23.

6. Agbabiaka LA, Madubuike FN, Uzoagba CU. Performance of catfish (*Clarias gariepinus*, Burchell, 1822) fed enzyme supplemented dried rumen digesta. Journal of Agricultural Biotechnology and Sustainable Development. 2012;4(2):22-26.

7. Adewolu MA, Ogunsanmi AO, Yunusa A. Studies on growth performance and feed utilization of two *Clarias* catfish and their hybrid reared under different culture systems. European Journal of Scientific Research. 2008;23(2):252-260.

8. Madu CT, Okwuego CC, Madu ID. Optimum dietary protein level for growth and gonadal maturation of female *Heterobranchus longifilis* (velenciennes 1840) brood stock. Journal of Aquatic Science. 2003;18(18):29-34.

9. Odedeyi DO. Survival and growth of hybrid (female) *Clarias gariepinus* (B) and male *Heterobranchus longifilis* fingerlings: Effect of broodstock sizes American-Eurasian. Journal of Scientific Research. 2007;2(1):19-23.

10. William HD, Ronaldo OC, Richard PS. (Broodstock Management: In Freshwater Prawn Culture: The Farming of *Macrobrachium rosenbergii* Blackwell Science Ltd 2000 edited by Michael Bernard New, Wagner Cotroni Valenti. 2007;443. APHA/AWWA/WPCF (American Public Health Association, American Water work Association, Water Pollution Control Federation). Standard methods for the examination of water and waste water, 16th edition, Washington, DC, American Public Health Association. 1995;1268.

11. Adewunmi AA. Catfish culture in Nigeria: Prospects and problems. Aquaculture Research. 2009;36:479-585.

13

Determinants of Access to Credit by Agribusiness Operators in the Kumasi Metropolis, Ghana

Stephen Opoku–Mensah[1*] and Hayford Agbekpornu[2]

[1]*Department of Agropreneurship, Institute of Entrepreneurship and Enterprise Development, Kumasi Polytechnic, Kumasi, Ghana.*
[2]*Mininstry of Fisheries and Aquaculture Development, Accra, Ghana.*

Authors' contributions

This work was carried out in collaboration between both authors. Author SOM designed the study, reviewed literature, collected data and wrote the first draft of the manuscript. Authors SOM and HA analyzed data as well as discussion of results and second draft of the manuscript. Authors SOM did the final write-up of manuscript. Both authors read and approved the final manuscript.

Editor(s):
(1) Yeamin Hossain, Department of Fisheries, Faculty of Agriculture, University of Rajshahi, Bangladesh.
(2) Anthony N. Rezitis, Department of Business Administration of Food and Agricultural Enterprises, University of Patras, Greece.
Reviewers:
(1) Anonymous, USA.
(2) Anonymous, France.

ABSTRACT

Aim: The study was conducted to analyze the factors that influence access to credit by Agribusiness operators in the Kumasi Metropolis of the Ashanti Region of Ghana.
Study Design: The study used a multi-stage sampling technique to select 151 SME operators engaged in the agribusiness sector. Well structured, mostly closed ended questionnaires were used to collect cross sectional data from the respondents.
Location and Place of Study: The study was conducted in the Kumasi Metropolis, which is the capital of Ghana's second largest city with a population of about 2 million people and has a relatively large agribusiness sector, especially in the informal sector.
Methodology: Both qualitative and quantitative methods were employed to analyze data. The Logit model was employed as the statistical tool to quantitatively analyze the factors that influenced access to credit by the agribusiness operators in the study area.
Results and Discussion: The empirical results showed that the factors that significantly influenced

---

*Corresponding author: E-mail: steopo@yahoo.com;

credit accessibility for respondents in the study area were the credit management skills, borrowing experience, possession of collateral security, firm size, extra income earned by operators and membership of business organization. The major constraining factors faced by respondents in their credit application from the formal sector include high interest rate, inadequate loan amount and unfavorable loan terms.

Conclusion and Recommendation: Agri-SME operators should be encouraged to form or join business organizations and also strengthen such associations for enhanced networking leverage. Agri–SME operators should be encouraged to build on their credit management skill by engaging more with financial institutions and strengthening relationship as a means of enhancing credit access. Financial institutions, especially the Rural and Community Banks and Savings and Loans Companies, should strive to offer more competitive terms and conditions, especially interest rates to Agri-SMEs that apply for credit as a group.

Keywords: Credit access; credit constraints; agribusinesses; agri-SMEs operators; Kumasi metropolis; logit regression.

1. INTRODUCTION

Agricultural and agribusiness finance is very significant in developing countries because it can potentially serve as an enabler of inclusive growth and poverty reduction, productivity enhancement, improved income for agribusinesses operators, and overall balanced regional development [1]. For a developing country like Ghana whose economy is fundamentally agriculture–driven, any financial investment in small-scale agribusinesses is very critical and worth considering because of their dominance and pervasiveness. Across the advanced world and developing countries, the small and medium enterprises (SMEs) sector, including Ghana's Agri-SMEs, generates the bulk of economic activity [2,3,4,5,6]. There is wide spread consensus that access to finance, particularly bank financing, can radically transform the outlook and performance of the SME sector, especially in the developing countries. For example, a number of studies including firm–level data from the World Bank over the years, show that inadequate financing is a greater obstacle for SME's than it is for large firms, particularly in the developing world, and that inaccessibility to finance constrains the growth of the SME sector more than that of large companies [7,8,9,10]. For many agribusinesses access to credit can readily support investments in productive operations, allow for the adoption of latest technologies and the scale-up of production activities to enhance productivity and competitiveness [11,12,13] and increase profitability [14].

Although support from development agencies, international donors and government to boost agriculture production in recent years through various policy interventions and programs in Ghana has increased (for example, Millennium Development Authority – MiDA program, Root and Tuber Improvement and Marketing Program - RTIMP, Northern Rural Growth Program – NRGP etc), lack of access to formal credit, especially for agribusinesses engaged in downstream processing, distribution and marketing has constrained the fullest realization of enhanced value-added and economic growth [1]. Agribusinesses and their value chain actors usually need a comprehensive suite of well-crafted financial services and products to improve their overall productivity and growth. Smallholder farmers need working capital to procure inputs and adopt modern technologies (improved seeds, fertilizer application, crop protection chemicals, basic storage facilities etc) that enhance output and profit. Aggregators, processors, and marketers of agricultural commodities also need credit to undertake product transformation and other downstream value-adding activities. Conceptually, the flow of adequate, well-structured and prompt credit facilities through these agri-value chains can potentially stimulate the much-needed growth within the agribusiness sector which remains largely rudimentary for a country like Ghana.

1.2 Problem Statement

It is rather ironic that in spite of the obvious economic growth potential of the agribusiness sector to the economy of developing countries including Ghana, agribusinesses are the most constrained when it comes to the supply of financial investments or access to finance [15,16,17]. Currently, only about 6% of total commercial bank credit goes to the agriculture sector, including cocoa, forestry and fisheries [18], and overall only 2%–10% of Ghana's financial institution's portfolios support agribusiness, which is much lower than agriculture's share of GDP (30%) in 2010. The formal financial sector particularly, commercial banks, which makes up the mainstream source of finance for businesses across countries (see [19,20,21,22]) are reluctant to offer credit to SMEs in general and the agriculture sector in particular. The formal financial sector is reluctant to lend to agribusinesses, including upstream primary production, because of the real and perceived high risk of the sector. Small agribusiness operators on the other hand also perceive formal credit as inaccessible due to rigid terms and conditions. For example the demand for more secured collateral, regular loan repayments typically on monthly basis instead of seasonal repayments and cumbersome loan applications and disbursement process are all deterrent to most agribusiness operators. These demand-side and supply-side constraints have resulted in a yawning financial gap within the formal credit market in Ghana. Increasingly therefore, many agribusiness entrepreneurs including input-suppliers, smallholder producers, distributors, processors, wholesalers and retailers usually resort to informal sources of credit to finance their businesses. Unfortunately these informal credit services come as unreliable, costly, inefficient and very expensive [23]. According to the International Finance Corporation (IFC), access to finance for agribusinesses is severe for upstream activities, particularly farm production than it is for downstream operators like processors, wholesalers, marketers, and retailers [24]. However, [25] has argued that access to credit for downstream Agri-SMEs is as challenging as it is for upstream production activities.

Given the crucial role of credit in enhancing the competitiveness of agribusinesses, especially in value-adding activities, this study attempts to gauge the credit situation for agribusinesses. This study attempts to find out whether agribusiness operators have easy access to credit. For those who are able to get access to credit, what are the major sources of credit? What factors influence access to credit by Agri-SMEs and what constraints do they meet in their credit acquisition process from the formal financial sector in Ghana? The goal of this paper was to investigate the factors that influence access to credit by agribusinesses in the study area. Specifically, the study sought

to examine the credit status and experience of agribusiness operators in the study area; to measure and quantify the factors that influence credit access and lastly, identify and rank the constraints in loan acquisition by agribusiness operators.

While empirical research on credit for SME's in general and smallholder farmers in particular abound in Ghana, there is a dearth of information on credit access for Agri–SMEs operating within the agribusiness sector. The outcome of this study will bridge the research gap, particularly credit experience and accessibility by Agri-SMEs along the pre-farm and post-farm end of the agricultural value chain.

2. LITERATURE REVIEW

Agricultural SME finance is defined as financial services for small and medium scale enterprises engaged in agricultural production and production related activities along the agribusiness value chain, such as input-supply, processing, wholesaling, marketing, trade, and export [24]. This study considers those Agri–enterprises engaged in off-farm activities within the agricultural value chain.

Credit is an instrument whose effectiveness depends on the economic and financial policies that go with it [26]. According to [27] credit means the ability to command capital in return for a promise to pay at some specific time in the future. Access to credit is very crucial to agribusiness, especially in less developed countries of the world. Access to credit (formal or informal) is the ability of an individual to borrow from a particular source, although for a variety of reasons he or she may choose not to [28]. The degree to which a person can get access to credit is measured by the credit limit of the individual. Thus people with positive credit limits are said to have access and vice-versa.

According to [29], 'credit accessibility' refers to the ease or difficulty of acquiring credit from borrowers to enhance business performance. Access to credit (also called financial inclusion) is the absence of both price and non-price barriers in the use of financial services. [30], therefore posit that access to credit is limited to a small proportion of the population who can overcome significant barriers to credit such as high minimum balance for account opening, collateral requirements and a long and costly bureaucratic process.

There are a plethora of arguments about financing for SME operators and the factors that enhance or constrain their access to or demand for credit. In the forgoing literature, this study examines some key variables that influence access to formal credit using evidence from SMEs in agribusiness as well as non-agriculture related activities. In other words, the study assumes that Agri–SMEs behave and thus share some similar characteristics or profile with both regular SMEs and agribusinesses.

2.1 Ghana's Financial Sector

Ghana's financial sector is classified into three: formal, semi-formal and informal. The formal sector includes Commercial Banks, Non-Bank Financial Institutions (NBFIs), Rural and Community Banks (RCBs), and Savings and Loans Companies (SLCs). As at 2010, Ghana had 23 major banks, including, 3 development banks, 4 commercial banks, and 16 universal banks. The NBFI sector includes 36 institutions made up of 17 savings companies, 13 savings and loans companies, 4 leasing companies, 1 discount house, and 1 mortgage company. The 23 major banks account for about 90% of the total assets of the banking and non-banking sector, whereas the NBFIs account for 5% of the total assets [31]. Most financial institutions in the formal sector are concentrated in the urban and peri-urban areas of Greater Accra, Ashanti and the Eastern regions, with limited outreach to the rural areas of Ghana. According to [18] the formal financial sector controls about 40% of the money supply in the overall financial sector with the remaining amount outside the formal system [32]. Thus, institutions such as RCBs and semi-formal and informal financial service providers play an important role in addressing the lack of access to financial services for many clients outside the formal sector.

2.2 Empirical Analysis and Evidence of Determinants of Access to Credit

Access to financial services, particularly credit, is dependent on or influenced by several factors which are within and beyond the control of borrowers. Based on evidence from several empirical findings, this study conveniently conceptualized that credit access by agribusinesses is influenced by four major factors namely socio-economic characteristics, managerial attributes, firm characteristics, and institutional factors.

2.2.1 Socio-economic characteristics and management attributes

Several demographic and socio-economic variables have been cited in various studies as determining access to credit by SME borrowers. [33] in a study of determinants of access to credit by SMEs in Kenya showed that the marginal effects of educational level, and occupation type (salaried, off-farm income, self-employment) significantly and positively influenced access to credit while distance to credit source and total value of income were negatively significant to credit access. However, age, gender, marital status and group membership did not significantly influence credit access. Females in many developing countries seem to have limited access to credit and are even discriminated against [34,35] compared to their male counterparts, apparently due to socio-cultural factors. However, studies by [36,37] in Ethiopia showed that the formal financial institutions offered more credit to female headed households than their male counterparts.

Also [38] in an analysis of determinants of access to institutional credit by SMEs in India showed that business registration, accounts record keeping, and higher education of business owners positively influenced access to credit. This is because well-educated entrepreneurs are better placed to have access to information on credit, have high financial literacy ability and good managerial skills in production, finance and marketing compared to their less educated or illiterate counterparts [39]. Thus there is a close relationship between the educational status of entrepreneur and their managerial competency and hence their propensity to easily get access to credit. Indeed empirical studies have also proven that managerial competencies of SME operators greatly influence ease of access to credit and vice–versa [40,41].

In another study in Limpopo, South Africa [42] found that loan repayment periods, distance between borrower and lender, and business experience of borrower were significant but negatively influenced credit access. [38], showed that access to credit by SMEs improved with business registration, record keeping and the educational level of the entrepreneur.

2.2.2 Firm–specific characteristics

For many Agri-SMEs as with mainstream SMEs the characteristics of the firm can influence or affect credit access. [43,44] in a study of enterprise access to credit in India, showed that the firm size, age, and collateral security of enterprise had positive and significant influence on access to bank loans. Financial institutions consider and are more likely to use firm size (proxied by employee numbers in some cases) as a criterion for lending [43]. This is because large firms are more diversified and less prone to failure [45] and are most often endowed with more valuable assets which enable them to attract long-term debt than smaller firms [46].

Firms and SMEs who have adequate fixed assets can use them as collateral security and this offers borrowers with high financial leverage to secure credit facilities [47]. Firms with more secured and valuable collateral therefore present better chances of accessing credit than their less endowed counterparts [48]. [42], again showed that assets accumulation or value of assets though significant, negatively influenced credit access in the study area.

The geographic proximity or distance between banks and customers has a relationship on a firm's use of leverage [49], because it allows banks to assess, evaluate and monitor clients at a cheaper cost and more efficiently.

The number of years a firm has been in operation also has an impact on access to credit. Thus Agri-SMEs with long years of operation would have developed or built business operational capacities and experiences including, management, financing, and marketing that are likely to make them more resilient and attractive to the formal financial sector compared to relatively new firms [50,51,52,53].

2.2.3 Financial institutions

Financial institutions which are the suppliers of financial services have peculiar characteristics that impact on the availability or otherwise of credit to borrowers. These institutional factors include a condition like operating an account with the financial institution which usually gives an indication of the credit management capacity of borrowers. Similarly the number of years a business owner or SME client has operated an account with the financial institution builds some level of relationship between lender and borrower, and thus enhances access to credit. The ability of a firm to keep basic and proper financial records, including financial statements, suggest some level of management capacity

thereby reducing information asymmetry problems which in turn also builds the level of transparency from the lender's perspective [54,55].

3. METHODOLOGY

3.1 Study Area and Location

The study was conducted in the Kumasi Metropolis, which is in the Ashanti Region of Ghana. The Ashanti Region lies between longitude 0º 15'W– 2º 15' and latitude 6º N– 7º 30'N of the equator. Kumasi is Ghana's second largest city and has a population of about 2,035,064, according to the 2010 National Housing and Population Census conducted by the Ghana Statistical Service (GSS). The city, a typical cosmopolitan area, depicts a fair share of vibrant economic activities, particularly commerce and artisanal activities within the informal sector. The metropolis is awash with thousands of small and medium enterprises engaged in several business activities of which agribusinesses and Agri–SMEs are well represented.

3.2 Sampling Technique and Data Collection

A multi-stage sampling technique was used to select some 151 respondents for the study. In particular agri–food processors, agri–food marketers, and agro-input dealers were purposively selected for the survey as they are quite dominant in the metropolis, particularly within the urban (central business district of Adum and the central market) and other peri-urban areas. The second stage involved a random selection of each respondent. Data were collected using well structured, mostly close ended questionnaires that captured information on the socio-economic and firm characteristics as well credit history and experience of the respondents.

3.3 Analytical Framework and Specification of Econometric Model

A decision by an entrepreneur to go for credit is influenced by a number of factors and considerations. Decision and choice model of this nature usually involves the use of an appropriate utility maximization approach such as the logistic regression model. The logistic regression model was employed to analyze the factors that influence an entrepreneur's ability to access credit from the formal financial sector. Logistic regression is particularly useful where the dependent variable (access to credit) is dichotomous.

In this analysis, Y_i represents an entrepreneur's decision to access or not to access credit from the formal financial institution. Y_i is assumed to be dependent on a vector of individual socio-economic characteristics, as well as institutional factors (X_i). The relationship between the dependent and explanatory variables is stated as:

$$Y_i = a + \beta' X_i + \mu_i \dots\dots\dots\dots\dots\dots\dots\dots (1)$$

Where Y_i (which is dichotomous) takes on the value of '1' or '0'.

The decision of the agribusiness operator to access credit is informed by the marginal cost and benefit he or she expects to derive from the use or otherwise of the credit. However, practically speaking the marginal utility (cost and benefit) is not observed, thus equation (1) cannot be estimated. The observation can only be derived from the responses made by the respondents in the survey. Thus, another variable is introduced to capture the reality of the use or non-use of credit as Y_i^*, such that:

$$Y_i^* = \begin{cases} 1= & \textit{if respondent was able to} \\ & \textit{access credit from formal sector} \\ 0= & \textit{if respondent did not} \\ & \textit{access credit from formal sector} \end{cases} \dots\dots \quad (2)$$

Thus

$$Y_i^* = \alpha + \beta^i X_i + \mu_i \dots\dots\dots\dots\dots\dots\dots\dots\dots\dots \quad (3)$$

Though in the analysis of data involving binary choice models, three commonest approaches such as the Linear Probability Model (LPM), the Logit and the Probit Models, are possible, several research analysts have settled for the latter two. The Logit and Probit models according to [56,57] are preferred because of their similarity and most importantly to avoid the limitations associated with the LPM, where the estimated probabilities can lie outside 0-1 [58].

According to [59] both the Logit and Probit are non-linear models and are estimated using maximum likelihood (ML) method. In addition, [60] noted that both Logit and Probit models guarantee that the estimated probabilities lie between the logical limit of 0 and 1. Due to these

advantages, the Logit and the Probit models are the most frequently used models when the dependent variable happens to be dichotomous [61,62,63]. The main difference between the Logit and Probit is in the nature of their distribution which is captured by cumulative distribution function (CDF). In this study, the logistic regression model is selected because it has been widely used by many researchers and most importantly because of its comparative simplicity, convenience, flexibility and also a powerful estimator of models [64].

Following [59,62,33], the cumulative logistic probability model is econometrically specified as:

$$P_i = F(Z_i) = F(a + \textstyle\sum_{i-1}^{n} \beta_i X_i) = \frac{1}{1+e^{-z_i}} \dots \quad (4)$$

Where,

$P_i =$ Is the probability that an individual accesses credit given X_i.

$X_i =$ Represents the ith explanatory variables

$e =$ Represents the base of natural logarithms, which is approximately equal to 2.718

$\alpha =$ Constant terms

$\beta_i =$ Coefficient of independent variables to be estimated

Central to the use of logistic regression is the Logit transformation of P_i given by Z. That is, to obtain linearity, the natural logarithms of odds ratio equation (4) is taken, which results in the Logit model as given by:

$$Z_i = In\left(\frac{P_i}{1-P_i}\right) = \alpha + \beta_i X_i + \beta_2 X_2 + \dots + \beta_n X_n \dots\dots\dots\dots\dots\dots\dots\dots\dots\dots\dots \quad (5)$$

Where Z_i is the indicator of the agribusiness entrepreneur's access to credit or not; P is the probability of the event's occurrence; and X_i is a vector of household socio-economic, managerial, firm characteristics, and factors characteristic of financial institutions.

3.4 Specification of Empirical Model

In this study, 'access to credit' refers to those respondents who were able to receive credit from a lending institution. This takes a dichotomous response variable of 'yes' or 'no' for those who had credit and those who did not respectively. Access to credit by agribusiness SMEs can be formulated implicitly as:

$$Agri - SME\ Credit\ Access = f \begin{pmatrix} socioeconomic\ characteristics, \\ management\ characteristics, \\ firm\ characteristics, Institutional\ factors \end{pmatrix}$$

The Logit Model can therefore be specified as:

$Credit\ access = \beta_o + \beta_1 Age + \beta_2 Gender + \beta_3 EducL + \beta_4 YBExp + \beta_5 OIncome + \beta_6 MBOrg + \beta_7 FirmSize + \beta_8 GMSales + \beta_9 Colattsec + \beta_{10} CrdtSkill + \beta_{11} MgtRec + \beta_{12} BRegist + \beta_{13} ProxFI + \beta_{14} CrdtExp + \mu_i$ .... (6)

A summary of variables and descriptions used in the regression analysis is presented in Table 1.

Table 1. Description of variables used in regression of access to credit by agri-SME's

Variable	Description	A priori sign
Age	Age of agribusiness entrepreneur (yrs)	+/-
Gender	Sex of respondent 1=male; 0=female	+
EducL	Educational status/years of schooling (yrs)	+
YBExp	Years of experience in business (yrs)	+
OIncome	Other source of income 1=yes; 0 = no	+
MBOrg	Membership of business association 1=yes; 0 = no	+
FirmSize	Firm size measured by number of hired employees	+
GMSales	Estimated Gross Monthly Sales (GH¢)	+
Colattsec	Availability of securable collateral 1 = yes; 0 = no	+
CrdtSkill	Credit management skill of respondent 1 = good ; 0 = not good	+
MgtRec	Ability to keep management records 1=yes; 0 = no	+
BRegist	Registration status of business 1 = registered ; 0 = otherwise	+
ProxFI	Proximity to financial institution 1 = close; 0 = far	+
CrdtExp	Years of banking or credit with FI measured in (yrs)	+
μ_i	Error term	

4. RESULTS AND DISCUSSION

4.1 Credit Experience, Status and History of Agri–SME Operators

The summary of credit experience and status of respondents is presented in Table 2. The result showed that the major credit need for most of the Agri–SME operators was for working capital (60%), business expansion (18%) and start-up capital (10%). This confirms the well documented assertion that for many SMEs, working capital for operational activities was the most critical credit need. All the Agri-SMEs in the survey operated as sole proprietors who typically require funds to take care of routine operational activities hence the high demand for working capital loans. All the respondents indicated that they had business relationship with or patronized at least one of the financial services or products offered by the financial institutions. Most of the Agri-SME operators either banked with the RCB–Rural and Community Banks (29%) or the SLC–Savings and Loans Company (26%). Also 23% dealt with the UB–Universal Banks while 7% banked with

Micro-Finance Companies. The relatively high preference for the services of RCBs and SLCs in a typical cosmopolitan area like Kumasi suggests and gives an indication of the growing importance of these financial institutions in the country as far as SME financing is concerned. Although all the 151 respondents had enjoyed some form of banking services, only about 55% stated that they had ever applied for a loan before, especially from the formal financial sector.

The results again revealed that contrary to common assertion, most respondents (67%) indicated that the financial institutions did not *strictly* demand collateral security while (33%) indicated that collateral security was required. These results are not surprising, given that a growing number of SLCs and RCBs in urban Ghana now place much emphasis on the relationship and bank transactional history of clients as opposed to the universal banks where collateral security is a major requirement.

Furthermore, when the respondents were asked to state whether they were able to meet loan application requirements apart from the usual demand for collateral security, 12% indicated that they were *'always'* able to, 67% stated *'sometimes'* and 21% stated *'not at all'*.

Credit management is important for many business operators who require especially formal credit. When respondents were asked to rate their skills in credit management, 58% rated that it was *good* while 42% stated that it was *not good*.

Finally, the survey showed that about half of the respondents (53%) declared that they had

access to credit while 47% stated otherwise. And for those who have ever attempted applying for credit, a little above average (54%) were of the opinion that loan application was cumbersome while 46% stated otherwise.

4.2 Logistic Analysis of Factors that Influence Access to Credit by Agri-SME Operators

The Logistic regression model was employed to analyze the factors that affect access to credit by Agri-SME operators in the Kumasi Metropolis.

Table 2. Summary of credit experience and status of agri-SME's in the metropolis

Variable	Frequency	Percentage (%)
Primary credit needs		
• Working capital	91	60
• Start-up capital	15	10
• Inventory/stocks	7	5
• Capital/assets	10	7
• Business expansion	28	18
Form of business status		
• Sole proprietorship	151	100
• Partnership	0	0
• Limited liability company	0	0
Source of financial services		
• Universal banks	34	23
• Savings and loans	39	26
• Rural and community banks	44	29
• Micro finance institutions	11	7
• Others	23	15
Credit application history		
• Yes	83	55
• No	67	45
Collateral demands from lenders		
• Yes	35	33
• No	70	67
Able to meet requirements		
• Always	11	12
• Sometimes – partly	61	67
• Rarely – not all at all	19	21
Loan application process		
• Simple	61	46
• Cumbersome	71	54
Credit management skills		
• Good	87	58
• Not good	64	42
Access to credit		
• Yes	80	53
• No	71	47

Source: Field survey, 2013

Results from the regression analysis, as posted in Table 3, indicate that the goodness of fit of the logistic model tested by means of the Log-Likelihood Ratio show a 1% level of significance. This means that the explanatory variables included in the model jointly explain the probability of Agri–SME operators' decision to access credit from financial institutions. Out of the 14 explanatory variables, six; namely, other income, membership of business organization, firm size, borrower/credit experience, collateral security, and credit management skills were found to be statistically significant determinants of credit accessibility. Membership of a business organization was significant and positively influenced access to credit at 10% level and this is in tandem with the findings of Kiplimo [33]. It is most likely that business organizations or associations that are well structured, functional, and vibrant are able to offer network opportunities, to link up members to credit sources, and thus enhance access to credit. Credit experience of Agri-SME operators also had a positive and significant influence on credit access at 1% level. Similarly credit management skill of Agri-SME operators and their ability to present valuable collateral security were significant and positively related to access to credit at 5% probability level. The marginal effects of the Logit model showed that the probability of credit access by Agri-SME operators increases by about 11%, 17% and 30% for every a percent increase in the credit experience, credit management skill and collateral respectively. All these three variables (credit experience of borrower, credit management skills and collateral security) and their outcomes in this study corroborate the findings of Fatoki and Assah [40], Herrington and Wood [41], Kohli [43], Abor and Quartey [48], and Eastwood and Kohli [44]. Formal financial institutions now place some level of premium on client relationship, credit and financial management skill of clients and their ability to provide valuable collateral security. These attributes or variables have always played an important role and enhanced access to credit from the formal sector, as shown in this study.

Expectations are that agribusiness operators who had other sources of income would be better placed to use credit from the formal financial institutions. However, results from the analysis showed that the variable rather had a negative relationship with the dependent variable. The possible reason for this could be that business operators earning extra income from other

activities may not experience any financial constrains as to warrant external financing, hence the negative impact on access to credit. Another reason could also be that the financial institutions did not rate this as a necessary condition. The study also revealed that firm size though significant at 10%, negatively influenced credit access by Agri-SME operators. This finding is contrary to the assertions of Kholi [43] and Burkhart and Ellingsen [46], who indicated that large firms have relatively easy access to credit than smaller firms. Indeed it is not surprising that this variable negatively influenced credit access as most of the Agri-SMEs had on average staff strength of three persons

On the other hand, none of the socio-economic variables included in the model had any significant impact or influenced Agri-SME operators' decision to get access to credit. Also coefficients of variables such as the gross monthly sales, years of business experience and proximity to financial institution were not statistically significant to influence access to credit; thus contrary to a priori expectation and some empirical evidence. It is most likely that although some financial institutions demand collateral, their decision to approve credit fell short of determining the value of these assets. Again it is most likely that the financial institutions did not consider or make lending decision based on gross monthly sales of loan applicants probably because those data could not be verified. Rather, the banks depended more on the relationships they had with their customers vis-à-vis borrowing experience and credit management skills.

4.3 Constraints in Loan Application Process for Agri–SME's in the Metropolis

Agribusiness operators, like most SMEs have peculiar constraints in the loan acquisition process. In this study, the respondents were asked to rank in order of importance or severity the most constraining factors that affect their credit or loan application with their bankers, mostly formal credit sources. Out of the total of 151 Agribusiness operators involved in the survey, between 74 and 111 responded to the various issues of credit constraints. The results as presented in Table 4 shows that about 88% of Agri-SME operators ranked *high interest rate* as the most severe constraint, 57% ranked *high interest rate* as the next constraint, whiles 50% ranked *inadequate amount of loans* as the third constraining factor. More than 50% of the respondents contend that there was a difference between the loan amount requested and the amount granted. Next a Kendall Coefficient of Concordance Test (W), to analyze the level of agreement among the agribusiness operators, showed that there was a significant difference at 1% level between the rankings of the constraints faced by respondents. The results of the analysis as presented in Table 5, revealed that *high interest rate* was ranked topmost with a mean rate of 1.47, followed by *inadequate loan amount* at 2.98 and *unfavorable loan term* with a mean score of 3.13. Lastly, the respondents perceived that the issue of cumbersome loan application procedures was the least severe constraint among the list.

Table 3. Logistic (marginal effect) regression results of the factors influencing credit access to agri-SME operators

Number of obs =102
LR chi2(14) = 67.11
Prob > chi2 =0.0000
Pseudo R^2 = 0.5356
Log likelihood = -29.088872

Variable	Marginal effect	Coef.	Std. Err.	Z	P>|z|	95% Conf. interval	
Age	-0.00142	-0.01546	0.039554	-0.39	0.696	-0.09298	0.06206
Gender	0.03635	0.39451	0.73425	0.54	0.591	-1.04459	1.83362
EducL	0.01515	0.16442	0.36795	0.45	0.655	-0.55676	0.88559
BRegist	-0.09513	-1.03256	0.98689	-1.05	0.295	-2.96682	0.90171
OIncome	*-0.13789*	*-1.49666*	*0.88048*	*-1.70*	*0.089**	*-3.22237*	*0.22906*
MBorg	*0.17524*	*1.90205*	*1.10792*	*1.72*	*0.086**	*-0.26944*	*4.07354*
YBexp	-0.00152	-0.01647	0.05726	-0.29	0.774	-0.12871	0.09577
Firmsize	*-0.03851*	*-0.41802*	*-0.23662*	*-1.77*	*0.077**	*-0.88175*	*0.04579*
GMsales	-0.00002	0.00021	-0.00018	-1.13	0.259	-0.00057	0.00015
MgtRec	-0.09626	-1.04475	0.84876	-1.23	0.218	-2.70830	0.61880
ProxiFi	0.05469	0.59363	0.91713	0.65	0.517	-1.20392	2.39118
CrdtExp	*0.10891*	*1.18206*	*0.35270*	*3.35*	*0.001***	*0.49078*	*1.87334*
Colattsec	*0.30174*	*3.27502*	*1.49707*	*2.19*	*0.029***	*0.34080*	*6.20923*
CredtSkill	*0.17189*	*1.94944*	*0.88610*	*2.20*	*0.028***	*0.21272*	*3.68616*
Const		-1.41213	1.65883	-0.85	0.395	-4.66337	1.83911

*Significant at 10%; ** Significant at 5%; and * ** Significant at 1%*

Table 4. Percentage distribution of constraints to loan application

Variable	Percentage ranking	Ranking
High Interest rate	88.30%	1
Unfavorable term of loan	57.00%	2
Inadequate loan amount	50.00%	3
Delayed loan disbursement	35.40%	4
Cumbersome procedures	25.50%	5

Source: Field survey, 2013

Table 5. Rankings of constraints Kendall's (W) in loan application process

Constraints	Mean rank
High interest rate	1.47
Unfavorable term of loan	3.13
Inadequate loan amount	2.98
Delayed loan disbursement	3.53
Cumbersome procedures	3.88

Test statistics: N=30, df = 4; $\chi2$ = 42.048; and P<0.000

5. CONCLUSION AND RECOMMENDATIONS

This study sought to analyze the factors that influence access to credit for Agri-SME operators in the Kumasi Metropolitan Area of Ghana. The study revealed that Agri-SMEs in the study area have a relatively good credit history or credit status, with all the respondents patronizing one form of financial services or the other, particularly from the Rural and Community Banks, Savings and Loans Companies and Universal Banks. The most important or prevalent credit need for most respondents was working capital loans, although only about 53% were able to get access to such credit. The logistic regression analysis of one hundred and fifty-one (151) respondents showed that credit access was influenced significantly by six variables namely; extra income earned by respondents, firm size, borrower experience, credit management skills, and possession of collateral security. Contrary to expectation, however, factors such as years of business experience of Agri-SME operators, proximity to financial institution and gross monthly sales were insignificant and did not influence access to credit.

Finally, the study revealed that factors that influence access to credit by Agri-SME operators were similar to those in the mainstream small business sector (or non-agricultural related businesses). This study recommends that Agri-SME operators build and improve upon their credit history and experience by banking with the formal sector for the variety of products and services they offer. For example, agribusiness enterprises with high daily or weekly sales turnover should lodge proceeds with their bankers to build their relationship and credit history, and thus enhance their chances of access to credit. Agri-SMEs should also be encouraged to join well-functioning business organizations or groups which can offer a platform for business networking and linkages to financial services. This would also lead to overall improvement of credit management skills. Financial institutions on the other hand, especially from the formal sector, should increase their engagement with the Agri-SME operators to improve relationship and credit experience of clients by employing innovative means for credit accessibility.

COMPETING INTERESTS

Authors have declared that no competing interests exist.

REFERENCES

1. Sharma M, Zhang J. Analysis of prospects for delivering agricultural finance for sustainable development, expanding agricultural market opportunities and promotion of disadvantaged small farmers and MSMEs: Workshop on Enhancing Exports' Competitiveness Though Value Chain Finance. Background paper series. 2012;1-7.

2. Wendel C, Harvey M. SME credit scoring: Key initiatives, Opportunities and issues. The World Bank Group. Access Finance. No. 10; 2006.

3. OECD (Organization for Economic Co-operation and Development). Measuring globalization: OECD economic globalization indicators paris: OECD Publishing; 2010.

4. OECD (Organization for Economic Co-operation and Development). Financing SMEs and Entrepreneurs, OECD, Observer Policy Brief; 2006.

5. OECD (Organization for Economic Co-operation and Development). OECD small and medium enterprise outlook. Paris, OECD; 2000.

6. Beck T, Demirgüç-Kunt A, Martinez Peria MS. Bank financing for SMEs: Evidence across countries and bank ownership types. Journal of Financial Services Research. 2010;(39):35-54.

7. Schiffer M, Weder B. Firm size and the business environment: Worldwide survey results. IFC Discussion Paper, 43; The World Bank, Washington DC; 2001.

8. IADB. Unlocking credit: The quest for deep and stable lending. The Johns Hopkins University Press; 2004.

9. Beck T, Demirguc-Kunt A, Martinez PM. Bank Financing for SMEs around the World: Drivers, obstacles, business models, and lending practices, World Bank Policy Research Working Paper 4785; 2008. The World Bank, Washington DC World Bank, various years. Enterprise analysis surveys. The World Bank, Washington DC.

Available:http://www.enterprisesurveys.org
/

10.   Beck T, Demirgüç-Kunt A, Laeven L, Maksimovic V. The determinants of financing obstacles. Journal of International Money and Finance. 2006;(25):932–952.

11.   Beck T, Demirgüç-Kunt A, Maksimovic V. Financial and legal constraints to firm growth: Does firm size matter? Journal of Finance. 2005;(60):137–177.

12.   UNCTAD: Issues concerning SMEs, Access to finance, Geneva: United Nations; 1995.

13.   UNCTAD: Survey of good practice in public private sector dialogue. Enterprise development services New York and Geneva United Nations; 2001.

14.   Utterwulghe S, Fall B, Ivanovic D. Public – private dialogue for specific sector: Agribusiness. Investment Climate Department, World Bank; 2013.

15.   World Bank: Ghana agribusiness indicators: Economic and work sector report no. 68163-GH; 2012.

16.   International Finance Corporation (IFC): Scaling-Up SME access to financial services in the developing world. Washington, DC; 2010.

17.   International Finance Corporation (IFC): Scaling up access to finance for agricultural SMEs: Policy Review and Recommendations Washington DC: IFC; 2011.

18.   Bank of Ghana. (BoG). Quarterly Bulletin; 2011.

19.   Nair A, Fissha A. Rural banking: The case of rural and community banks in Ghana. Washington DC, 2010: The World Bank. Accessed from on February 23, 2011. Available:Http://siteresources.worldbank.or g/INTARD/Resources/GhanaRCBs web.p df

20.   Hallberg K. A market–oriented strategy for small and medium scale enterprises. International Finance Corporation Discussion Paper, 40; 2000.

21.   Beck T, Demirgüç-Kunt A, Maksimovic V. Financing patterns around the world: Are small firms different? Journal of Financial Economics. 2008;(89):467-87.

22.   Demirguc-Kunt A, Maksimovic V, Beck T, Laeven L. The determinant of financing obstacles. International Journal of Money and Finance. 2006;(25):932-952.

23.   Olomola AS. Effects of membership homogeneity on the design and performance of informal finance groups in Rural Nigeria. A Research Report Submitted to the African Economic Research Consortium (AERC) Nairobi, Kenya; 2000.

24.   IFC: Innovative agricultural sme finance models; 2010.

25.   World Bank. 2012. Global Financial Inclusion (Global Findex) Database. Avaliable:http://www.worldbank.org/globalfi ndex

26.   Nwaru JC, Determinants of arm and off-farm incomes and savings of food crop farmers in Imo State, Nigeria: Implication for poverty alleviation. Nigerian Agricultural Journal. 2005;36-42.

27.   David TJ. The business of farming, Published by Macmillan Education Limited; 1990.

28.   Diagne A, Zeller M. Access to credit and its impact on welfare in Malawi, Research Report, International Food Policy Research Institute, Washington, DC. 2001;116.

29.   Salahuddin A. Pertinent issues on SME finance in Bangladesh, The financial Express. 2006;1:589.

30.   Okurut N,Schoombe E, Servaas V. Credit demand and credit rationing in the informal financial sector in Uganda. Paper presented in Africa Development and Poverty Reduction: The Macro- Micro Linkage Forum; 2004.

31.   World Bank, Rural Banking: The case of Rural and Community Banks in Ghana; 2010. Avaliable:http://siteresources.worldbank.or g

32.   International Fund for Agricultural Development (IFAD). The republic of Ghana rural and agricultural finance program; 2008.

33.   Kiplimo JC. Determinants of access to credit by smallholder farmers in Eastern and Western Kenya. A master degree dissertation, Strathmore School of business, Strathmore University, Nairobi, Kenya; 2013.

34.   Buvinic M, Sebstad J, Zeidenstein S. Credit for rural women: Some facts and

lessons. Washington DC: International Center for Research on Women; 1979.

35. Morris GA, Meyer RL. Women and financial services in developing countries: A review of the literature. Economics and Sociology. Occasional Paper. 1993;2056.

36. Kedir A. Determinants of access to credit and loan amount: Household-level evidence from urban Ethiopia. Working Paper. 2007;7:3.

37. Mohamed K. Access to formal and quasi-formal credit by smallholder farmers and artisanal fishermen: A case study of Zanzibar. Tanzania: Mkuki na Nyota Publishers; 2003. ISBN 9987-686-75-3.

38. Nikkaido Y, Jesim P, Mandira S. Determinants of access to institutional credit for small enterprises in India; 2010.

39. Kumar A, Fransisco M. Enterprise size, financing patterns, and credit constraints in Brazil: Analysis of data from the investment climate assessment survey, World Bank Working Paper, 49; 2005.

40. Fatoki O, Asah F. The impact of firm and entrepreneurial characteristics on access to debt finance by SMEs in King Williams' Town, South Africa. International Journal of Business and Management. 2011;6(8).

41. Herrington M, Wood E. Global Entrepreneurship Monitor, South African Report; 2003. May 5, 2008. Available:http://www.gbs.nct.ac.za/gbsweb b/userfiles/gemsouthafrica2000pdf

42. Chauke PM, Motlhatlhana ML, Pfumayaramba TK, Anim FDK: Factors influencing access to credit: A case study of smallholder farmers in the Capricorn district of South Africa. African Journal of Agricultural Research. 2013;8(7):582-585.

43. Kohli R. Credit availability and small firms: A probit analysis of panel data. Reserve Bank of India Occasional Papers. 1997;18(1).

44. Eastwood R, Kohli R. Directed credit and investment in small scale industry in India: Evidence from firm-level data 1965-78. Journal of Development Studies. 1999;35(4).

45. Honhyan, Y. The determinants of capital structure of the SMEs: An empirical study of Chinese listed manufacturing companies; 2009.

46. Burkart MC, Ellingsen T. In-kind finance: A theory of trade credit. American Economic Review. 2004;94(3):569-590.

47. Barbosa EG, Moraes CC. Determinants of the firm's capital structure: The case of the very small enterprises; 2004 [Online]. (March 12, 2012) Available:http://econpa.wustl.edu.8089/eps /fin/papers 0302/0302001.pdf

48. Abor J, Quartey P. Issues in SME Development in Ghana and South Africa. International Research Journal of Finance and Economics. 2010;(39):218–228.

49. Berger A, Udell G. Small business credit availability and relationship lending: The importance of bank organizational structure, Economic Journal. 2002;1-36.

50. Chandler JG. Marketing tactics of selected small firms in the East London CBD area. South Africa: University of South Africa; 2009. (August 2, 2012). Available:http://uir.unisa.ac.za/xmlui/handl e/10500/1878

51. Klapper L, Laeven L, Rajan R. Entry regulation as a barrier to entrepreneurship. Journal of Financial Economics. 2010;82(3):591-623.

52. Ngoc TB, Le T, Nguyen TB. The impact of networking on bank financing: The case of small and medium enterprises in Vietnam. Entrepreneurship Theory and Practice. 2009;33(4):867-887.

53. Bougheas S, Mizen P, Yalcin C. Access to external finance: Theory and evidence on the impact of monetary policy and firm-specific characteristics. Journal of Banking & Finance. 2005;30(1):199-227.

54. Kitindi EG, Magembe BAS, Sethibe A. Lending decision making and financial information: The usefulness of corporate annual reports to lender in Botswana. International Journal of Applied Economics and Finance. 2007;1(2):55-60.

55. Sarapaivanich N, Kotey B. The effect of financial information quality on ability to access external finance and performance of SMEs in Thailand. Journal of Enterprising Culture. 2006;14(3):219-239.

56. Amemiya T. Adavanced Econometrics. T.J Press Pad Stow Ltd: Great Britain; 1981.

57. Gujarati DN. Basic Econometrics 4th edn, McGraw-Hill Companies; 2004.

58. Pindyck RS, Rubinfeld DL. Econometric Models and Economic Forecasts. (2nd ed). McGraw- Hill Book Co. New York; 1981.

59. Brooks C. Introductory Econometric for Finance. Cambridge; Cambridge University. Press; 2008.

60. Wooldridge JM. Introductory Econometrics: A Modern Approach (4th ed.);. Cengage Learning. Tsinghua University Press, People's Republic of China; 2009.

61. Liao T. Interpreting probability models: Logit, probit, and other generalized linear models. Thousand Oaks, CA: Sage; 1994.

62. Maddalla GS. Limited dependent and quantitative variables in econometrics. Cambridge: Cambridge University Press; 2001.

63. Gujarati DN. Basic econometrics 4th ed. McGraw-hill companies; 2004.

64. Sirak M, Rice JC. Logistic regression: An introduction. In B. Thompson, ed., Advances in Social Science Methodology, Greenwich, CT: JAI Press. 1994;(3):191-245.

14

Haematology and Carcass Visual Appraisal of Broiler Chickens fed Supplemental Diets of *Aspilia africana*, *Azadirachta indica* and *Centrosema pubescence* Leaf Meals in Humid Tropical Nigeria

B. B. Okafor[1], G. A. Kalio[1*], H. A. Manilla[1] and O. N. Wariboko[1]

[1]*Department of Agricultural Science, Ignatius Ajuru University of Education, Ndele Campus, P.M.B. 5047, Port Harcourt, Nigeria.*

Authors' contributions

This work was carried out in collaboration between all the authors. Author BBO collected and carried out the laboratory analysis of the blood samples. Author GAK designed the study, wrote the protocol and wrote the first draft of the manuscript. Author HAM supervised and thoroughly proof read the manuscript. Author ONW reviewed the experimental design and performed the statistical analysis. All authors read and approved the final manuscript.

Editor(s):
(1) Anonymous.
Reviewers:
(1) Aliyu Abdullahi Mohammed, Usmanu Danfodiyo University, Sokoto, Nigeria.
(2) Dieumou Felix Eboue, Department of Animal Production Technology, Catholic University of Cameroon, Cameroon.

ABSTRACT

A study to determine the haematology and carcass visual appraisal of broiler chickens fed basal feeds supplemented with different leaf meals was conducted. Four treatments: Basal proprietary broiler feed only (T_1 - PBF) as control, basal proprietary broiler feeds with *Centrosema pubescence* (T_2 - PBF + CLM), *Azadirachta indica* (T_3 - PBF + NLM) and *Aspilia africana* (T_2 - PBF + ASLM) respectively, were used in a completely randomized design (CRD). On the last day of a 63-day feeding and growth trial, a set of 2 ml blood samples were taken from 3 broilers per treatment into plastic tubes containing the anti-coagulant ethylene diamine tetraacetic acid (EDTA) for the determination of haematological parameters: PCV, Hb, RBC and WBC. The MCHC, MCH and MCV were also determined. Visual appraisals of their external body parts per treatment were also carried out. Results on the blood parameters of broilers fed *Aspilia africana*, *Azadirachta indica* and

---

Corresponding author: E-mail: ag.kalio@yahoo.com;

Centrosema pubescence leaf meals showed normal blood values recommended for healthy birds. Similarly, broilers fed these leaf meals showed a better appeal for their carcasses because of the yellow pigmentation of their body parts (shank, skin, beak and ear lobes). This will be an advantage to the consumers because it supplies vitamin A necessary for better vision. It was concluded that poultry farmers incorporate *Aspilia africana*, *Azadirachta indica* and *Centrosema pubescence* at 5% inclusion levels in broiler feeds because it is not deleterious and can be of additional advantage due to the attractiveness of their carcasses to consumers.

Keywords: Broilers; leaf meals; haematology; carcass body parts; visual appraisal.

1. INTRODUCTION

The feed crisis facing the poultry industry in Nigeria strongly suggests the need to investigate and utilize cheap and easily obtainable non-conventional feed resources. One of such non-conventional feed resources is leaf meal. The incorporation of protein from leaf sources in diets for broilers is fast gaining grounds because of its availability, abundance and relatively reduced cost. It had earlier been observed that leaf meal do not only serve as protein sources but also provide some necessary vitamins, minerals and also oxy-carotenoids which causes yellow colour of broiler skin, shank and egg yolk [1].

Many feed materials are fed to animals usually without recourse to their health and physiological implications on the animals. The commonest method for measuring these implications is through the haematology of the animals [2]. There has been the emphasis that nutritional studies should not be limited to performance and carcass quality alone, but its effect on blood constituent is also very relevant [3]. Therefore, the comparison of an animal's haematological and biochemical values with a reference interval will provide evidence for numerous conditions such as infection, malnutrition and stress [4]. Similarly, laboratory tests on blood have been revealed as very vital tools to detect any deviation from the normal in the animal or human body [5]. Furthermore, with the growing knowledge of the quality of animal protein supplies in human diets, and the preference for carcasses of good visual appraisal, it becomes imperative for poultry feed manufacturers and farmers to source tangible alternatives to cope with the current challenges in the meat market.

Plant leaves such as *Aspilia africana*, *Azadirachta indica* (Neem) and *Centrosema pubescence* abound within the humid tropical environment of the South-south geopolitical zone of Nigeria and has not been maximally utilized as leaf meal supplements in poultry feeds. However, investigations on the use of the leaves especially *Aspilia africana* and *Azadirachta indica* (Neem) has been on their utilization as traditional medicines in maintaining the health and welfare of both rural and urban dwellers in developing countries [6,7]. There is therefore the need to investigate the effects of predominant plant materials: *Aspilia africana*, *Azadirachta indica* (Neem) and *Centrosema pubescence* as leaf meal supplements on the haematology and carcass appearance of broiler chickens in the area.

2. MATERIALS AND METHODS

2.1 Experimental Site

The experiment was conducted at the poultry unit of the Teaching and Research Farm, Ignatius Ajuru University of Education Ndele Campus, Rivers State (Latitude 4°58' N and Longitude 6° 48' E), Nigeria.

2.2 Processing of Test Ingredients

Fresh leaves from the young stems of the test ingredients: *Centrosema pubescence*, *Aspilia africana* and *Azadirachta indica* (Neem) were manually harvested from the bush and fallow sections within the premises of the Ignatius Ajuru University of Education at Ndele Campus. The fresh leaves were collected in batches in labeled bags. The leaves together with their stems were spread on black polythene on an open floor in a greenhouse and air dried for seven (7) days until it became crispy and shredded to separate the leaves from the stems. Air drying of the leaves was carried out to reduce moisture content, to prevent fungal growth and easy milling of the materials. The milling was done with a hand grinding machine. The milled products represent the test materials which were incorporated differently in the proprietary basal concentrate feeds.

2.3 Birds and Distribution to Treatment Groups

A total of one hundred and eighty (180) day old of Marshal strain of broiler chicks were brooded in the brooding unit (deep litter system) for 4 weeks (28 days) using stove and kerosene lanterns. The chicks were randomly allotted to four (4) dietary feed treatment groups: basal proprietary broiler feed only (T_1 - PBF) as control, basal proprietary broiler feeds supplemented with *Centrosema pubescence* (T_2 - PBF + CLM), *Azadirachta indica* (Neem) (T_3 - PBF + NLM) and *Aspilia africana* (T_4 - PBF + ASLM). The respective leaf meals were incorporated at 5% supplemental levels into the basal proprietary broiler feeds.

2.4 Determination of Haematological Parameters

On the last day of a 63-day (9 weeks) experimental feeding and growth trial, a set of blood samples were taken from 3 birds per dietary treatment randomly. The blood samples were collected by carefully cutting the jugular veins of each birds and collecting 2 ml samples of blood into plastic tubes containing the anti-coagulant ethylene diamine tetraacetic acid (EDTA) for the determination of haematological parameters.

The Haematological values of the blood samples were estimated for packed cell volume (PCV) and haemoglobin (Hb) concentration. Red blood cell (RBC) and total white blood cell (WBC) as well as the differential WBC counts were determined using the Neubauer haemocytometer after appropriate dilution. Values for the constants: mean corpuscular haemoglobin concentration (MCHC), mean corpuscular haemoglobin (MCH) and mean corpuscular volume (MCV) were calculated from RBC, Hb and PCV values.

2.5 Visual Appraisal on Pigmentation of External Body Parts

At the end of the experiment after 63-days (9 weeks), the broilers from the leaf meal treatment groups (T_2, T_3 and T_4) and control (T_1 – without leaf meal treatment) were also physically assessed based on the yellowish pigmentation of the skin, beak, earlobe and shank.

2.6 Statistical Analysis

The haematological data obtained from birds in each treatment group were compared statistically on the basis of the different dietary treatments using Analysis of Variance (ANOVA) procedure for a Completely Randomized Design (CRD) [4]. Significant means were separated using the Duncan's New Multiple Range Test [8].

3. RESULTS AND DISCUSSION

3.1 Nutrient Composition of Leaf Meals and Basal Concentrate Broiler Finisher Feeds

The nutrient composition of the different leaf meals, the proprietary broiler feed (finisher) and those supplemented with the different leaf meals fed to the experimental broiler chickens is similar to those reported by [9].

3.2 Haematological Characteristics of Broilers

Table 1 shows results of the haematological characteristics of broiler chickens exposed to the experimental diets: T_1- non- supplementation with leaf meal) used as control and other feeds supplemented with different leaf meals: T_2 (proprietary finisher basal feed supplemented with *Centrosema* leaf meal – PFBF + CLM), T_3 (proprietary finisher basal feed supplemented with Neem leaf meal - PFBF + NLM) and T_4 (proprietary finisher basal feed supplemented with *Aspilia* leaf meal - PFBF + ASLM) for 5 weeks after brooding (finisher phase).

The packed cell volume (PCV) value of the broilers fed with the experimental diets T_1 (leaf meal un-supplemented feed) and those supplemented with different leaf meal: T_2 (proprietary finisher basal feed supplemented with *Centrosema* leaf meal – PFBF + CLM), T_3 (proprietary finisher basal feed supplemented with Neem leaf meal - PFBF + NLM) and T_4 (proprietary finisher basal feed supplemented with *Aspilia* leaf meal - PFBF + ASLM) were not significantly ($P> 0.05$) different. The PCV value of the broilers ranged from 31.00 – 33.33% for T_2 and T_1 respectively. The PCV values reported in the study for the broiler chickens were within the normal range (22 – 35%) reported by [10] for normal or healthy chickens. Similarly the values of PCV observed in this study for birds whose feeds were supplemented with Neem leaf meal (T_3) is within the range (30 – 40%) for broilers as reported by [1]. This implies that supplementation of broiler feeds with the different leaf meals has no deleterious effect on the broilers as they maintain their normal blood count [11].

The haemoglobin (Hb) value of the broilers fed with the experimental diets T_1 (leaf meal un-supplemented feed) and those supplemented with different leaf meal: T_2 (proprietary finisher basal feed supplemented with *Centrosema* leaf meal – PFBF + CLM), T_3 (proprietary finisher basal feed supplemented with Neem leaf meal - PFBF + NLM) and T_4 (proprietary finisher basal feed supplemented with *Aspilia* leaf meal - PFBF + ASLM) were not significantly ($P> 0.05$) different. The haemoglobin value of the broilers ranged from 9.97 – 11.07 g/dl for T_4 and T_1 respectively. The haemoglobin values reported in the study for the broiler chickens were within the normal range (7.0 – 13.0 g/d) reported by [6] for normal or healthy chickens. This implies that supplementation of broiler feeds with the different leaf meals has no deleterious effect on the broilers [11].

The Red blood cells (RBCs) value of the broilers fed with the experimental diets T_1 (leaf meal un-supplemented feed) and those supplemented with different leaf meal: T_2 (proprietary finisher basal feed supplemented with *Centrosema* leaf meal – PFBF + CLM), T_3 (proprietary finisher basal feed supplemented with Neem leaf meal - PFBF + NLM) and T_4 (proprietary finisher basal feed supplemented with *Aspilia* leaf meal - PFBF + ASLM) were not significantly ($P> 0.05$) different. The Red blood cells (RBCs) value of the broilers ranged from 3.40 – 3.70 x 10^{12}/l for T_4 and T_1 respectively. The Red blood cells (RBCs) values reported in the study for the broiler chickens were within the normal range (2.5 – 3.5 x10^{12}/l) reported by [10] for normal or healthy chickens. This implies that supplementation of broiler feeds with the different leaf meals has no deleterious effect on the broilers as they maintain their normal blood count [11].

The Mean corpuscular haemoglobin concentration (MCHC) value of the broilers fed with the experimental diets T_1 (leaf meal un-supplemented feed) and those supplemented with different leaf meal: T_2 (proprietary finisher basal feed supplemented with *Centrosema* leaf meal – PFBF + CLM), T_3 (proprietary finisher basal feed supplemented with Neem leaf meal - PFBF + NLM) and T_4 (proprietary finisher basal feed supplemented with *Aspilia* leaf meal - PFBF + ASLM) were not significantly ($P> 0.05$) different. The Mean corpuscular haemoglobin concentration (MCHC) value of the broilers ranged from 32.13 – 33.20 g/100 ml for T_4 and T_2 respectively. The Red blood cells (RBCs) values

reported in the study for the broiler chickens were within the normal range (26.0 – 35.0 g/100 ml) reported by [10] for normal or healthy chickens. This implies that supplementation of broiler feeds with the different leaf meals has no deleterious effect on the broilers as they maintain their normal blood MCHC [11].

The Mean corpuscular haemoglobin (MCH) value of the broilers fed with the experimental diets T_1 (leaf meal un-supplemented feed) and those supplemented with different leaf meal: T_2 (proprietary finisher basal feed supplemented with *Centrosema* leaf meal – PFBF + CLM), T_3 (proprietary finisher basal feed supplemented with Neem leaf meal - PFBF + NLM) and T_4 (proprietary finisher basal feed supplemented with *Aspilia* leaf meal - PFBF + ASLM) were not significantly ($P> 0.05$) different. The Mean corpuscular haemoglobin (MCH) value of the broilers ranged from 28.47 – 30.83pg for T_4 and T_2 respectively. The Mean corpuscular haemoglobin (MCH) values reported in the study for the broiler chickens were lower than the normal range (33.0 – 47.0 pg) reported by [10] for normal or healthy chickens. This variation may be due to difference in the breeds of fowls [10].

The Mean corpuscular volume (MCV) value of the broilers fed with the experimental diets T_1 (leaf meal un-supplemented feed) and those supplemented with different leaf meal: T_2 (proprietary finisher basal feed supplemented with *Centrosema* leaf meal – PFBF + CLM), T_3 (proprietary finisher basal feed supplemented with Neem leaf meal - PFBF + NLM) and T_4 (proprietary finisher basal feed supplemented with *Aspilia* leaf meal - PFBF + ASLM) were not significantly ($P> 0.05$) different. The Mean corpuscular volume (MCV) value of the broilers ranged from 88.50 – 91.17fl for T_4 and T_2 respectively. However, although broilers in the T_1 and T_2 exhibited higher MCV values, those of T_3 and T_1 were still within the normal values for birds. No mortality was recorded; hence the experimental diets were not deleterious.

3.3 Pigmentation Pattern of External Parts of Broilers

Table 2 shows the pattern of pigmentation or colouration (white, fairly yellow or very yellow) of the external body parts (shank, skin, beak and ear lobe) of broiler chickens exposed to their basal feeds (non-supplemented/supplemented) with different leaf meals. Results from the visual

Table 1. Haematology of broiler chickens fed basal concentrate diets supplemented with different leaf meals

Parameters	Treatments					
	T1(PFBF)	T2 (PFBF + CLM)	T3 (PFBF + NLM)	T4 (PFBF + ASLM)	Mean	±SEM
PCV (%)	33.33	31.00	33.00	30.33	31.92	2.88
Hb(g/dl)	11.07	10.50	10.97	9.97	10.63	0.85
RBC X10^{12}/l	3.70	3.40	3.70	3.43	3.56	0.35
MCHC (g/100 ml)	33.17	33.13	33.20	32.13	32.91	1.04
MCH (pg)	29.90	30.83	29.70	28.47	29.45	1.17
MCV (fl)	90.13	91.17	89.27	88.50	89.77	0.90

PFBF = Proprietary broiler Finisher Basal Feed; NLM = Neem Leaf Meal; CLM = Centrosema Leaf Meal; ASLM = Aspilia Leaf Meal; PCV = Packed cell volume; Hb = Haemoglobin; RBC = Red blood cell; MCHC = Mean corpuscular haemoglobin concentration; MCH = Mean corpuscular haemoglobin; MCV = Mean corpuscular volume

Table 2. Visual appraisal of pigmentation of body parts of broiler chickens fed basal concentrate diets supplemented with different leaf meals

Treatment	Colour of external body parts											
	Shank			Skin			Beak			Ear lobe		
	W	FY	VY	W	FY	VY	W	FY	VY	W	FY	VY
T$_1$(PBFF only	Yes	No	No	Yes	No	No	Yes	No	No	Yes	No	No
T$_2$ (PBFF + CLM	No	No	Yes	No	No	Yes	No	No	Yes	No	No	Yes
T$_3$ (PBFF + NLM)	No	Yes	No	No	Yes	No	No	Yes	No	No	Yes	No
T$_4$ (PBFF + ASLM)	No	Yes	No	No	Yes	No	No	Yes	No	No	Yes	No

PBFF = Proprietary broiler Finisher Basal Feed; NLM = Neem Leaf Meal; CLM = Centrosema Leaf Meal; ASLM = Aspilia Leaf Meal; W = White; FY = Fairly yellow; VY = Very yellow

appraisal revealed variations in the colour patterns of the treatment groups. It was observed that the leaf meal supplemented groups (T$_2$ = PFBF + CLM, T$_3$ = PFBF + NLM and T$_4$ = PBFF + ASLM) possessed yellowish pigmentations on their shanks, skin, beak and ear lobes. However, the degree of yellowness of these body parts slightly differed. Thus the T$_2$ (PFBF + CLM) treatment group showed very yellow (VY) colouration as compared to their fairly yellow (FY) (T$_3$ = PFBF + NLM and T$_4$ = PFBF + ASLM) counterparts. The body colouration of broilers in T$_1$ (PBFF only) showed white (W) colouration of the parts. The yellow pigmentation of the various leaf meal supplementation groups may be as a result of the presence some pigmenting agents in the different leaves [12]. This is can be attributed to the presence of different xanthophylls of the general family of carotenoids resulting in pigmentation of the skin, fat, breast, and shanks and egg yolk [1].

4. CONCLUSION AND RECOMMENDATIONS

The investigation on the blood parameters of broiler chickens fed leaf meal supplemented

diets of Goat weed (Aspilia africana), Neem (Azadirachta indica) and Centrosema pubescence showed normal blood values as recommended for healthy birds. This further confirms that the use of the leaf meals at 5% inclusion level in basal broiler feeds is not deleterious to birds. Similarly, it was revealed that broilers whose feed were supplemented with different leaf meals showed a better appeal of their carcasses for consumption. The carcasses of the leaf meal treatment groups were better because of the yellow pigmentation of the different body parts (shank, skin, beak and ear lobes) due to the presence of carotenoids in these leaves. This will be of a better advantage to the consumers because of the supply of vitamin A that is good for better vision. Poultry farmers can also incorporate them in their broiler chicken feeds as it can be of an additional advantage due to the attractiveness of their carcasses to consumers to improve their profit margins.

ACKNOWLEDGEMENTS

The authors appreciate the assistance of Mrs. Esther. N. Nlegwu, Mrs. Ayibakuro Sambo, Mrs.

Barigbuge Richard, Mrs. Mercy Amadi and Mr. Ezemonye Wome for the collection of data during and after the conduct of the feeding trials. The assistance of Mr. E.O. Ekpenyong and Mr. Ofem of the Soil Science and Animal Science Laboratories at the University of Calabar, Calabar, Nigeria for the analysis of the various feed samples is also most appreciated.

COMPETING INTERESTS

Authors have declared that no competing interests exist.

REFERENCES

1. Esonu BO, Opara MN, Okoli IC. Obikaonu HO. Udedibie C, Iheshiulor OOM. 'Physiological responses of laying birds to Neem (*Azadirachta indica*) leaf meal based diets, body weight, organ characteristics and hematology'. Online Journal of Health and Allied Science. 2006;5(2). Available:http://cogprints.org/5168/1/2006-2-4.pdf

2. Aro SO, Ogunwale FF, Falade OA. Blood viscosity of finisher cockerel fed dietary inclusions of fermented cassava tuber wastes. Proceedings of the 18th Annual Conf. of Anim. Sci. Assoc. of Nig. 2013; 74-77.

3. Animashaun RA, Omoikhoje SO, Bamgbose AM. Haematological and Biochemical Indices of Weaner Rabbits Fed Concentrate and Syndrella noniflora Forage Supplement. Proceedings of 11th Annual Conference of the Animal Science Association of Nigeria (ASAN), Ibadan, Nigeria. 2006;29–31.

4. Clifford GR, Briggs H. Hematologic and Biochemical Reference Intervals for Mountain Goats (*Oreamnos americanus*): Effects of Capture Conditions. North West Science. 2007;81(3):206–214.

5. Alemede IC, Adama JY, Ogunbajo SA, Abdullahi J. Haematological parameters of Savanna brown does fed varying dietary levels of flamboyant tree seed meal. Pakistan Journal of Nutrition. 2010;9(2): 167-170.

6. Adeniyi BA, Odufowora RO. *In vitro* Antimicrobial Properties of *Aspilia africana* (Compositae). African Journal of Biomedical Research. 2000;3:167-170.

7. Iwu MM. Handbook of African Medicinal Plants. CRP Press, Boca Raton Florida; 1993.

8. Obi IU. Statistical Method of Detecting between Treatment Means. SNAAP Press Nigeria Limited, Enugu, Nigeria; 2001.

9. Kalio GA, Manilla HA, Wariboko ON, Okafor BB. Mineral profiling, carcass quality and sensory evaluation of broiler chicken fed basal proprietary diets supplemented with leaf Meals. African Journal of Livestock Extension. 2014;14: 77–83.

10. Benerjee GC. A textbook of animal husbandry, 8th Edition. Oxford and IBH Pub., Co., PVT, LTD; 2004.

11. Sembulingam K, Sembulingam P. Essentials of medical physiology. 2nd Edition. Jaypee Brothers Medical Publishers (P) Ltd, New Delhi. 2002;840.

12. Devendra C. Shrubs and tree fodders for farm animals. Proceedings of a workshop in Denpasar, Indonesia. 1989;61–75.

15

Media Effects on Emergence and Growth of Moringa (*Moringa oleifera Lam*) Seedlings in the Nursery

A. E. Ede[1], U. M. Ndubuaku[1*] and K. P. Baiyeri[1]

[1]*Department of Crop Science, Faculty of Agriculture, University of Nigeria, Nsukka, Nigeria.*

Authors' contributions

This work was carried out in collaboration between all authors. Authors UMN and KPB designed the study. Author UMN wrote the protocol and the first draft of the manuscript. Author UMN reviewed the experimental design and all drafts of the manuscript. Author AEE performed the statistical analysis. All the authors managed the analyses of the study, identified the plants, read and approved the final manuscript.

Editor(s):
(1) Lanzhuang Chen, Laboratory of Plant Biotechnology, Faculty of Environment and Horticulture, Minami Kyushu University, Miyazaki, Japan.
(2) Velu Rasiah, Department of Environment and Resource Management, PO Box 156, Mareeba, QLD 4880, Australia And School of Science, Information Technology & Engineering, University of Ballarat, Australia.
(3) Craig Ramsey, United States Department of Agriculture, Animal & Plant Health Inspection Service, Plant Protection and Quarantine, Center for Plant Health Science and Technology, 2301 Research Blvd., Suite 108, Fort Collins, CO 80526, USA.
(4) Daniele De Wrachien, State University of Milan, Italy.
Reviewers:
(1) Anonymous, USA.
(2) Anonymous, Togo.
(3) Anonymous, Nigeria.
(4) Anonymous, Brazil.

ABSTRACT

This study was carried out in the Teaching and Research Farm of the Department of Crop Science, University of Nigeria, Nsukka to determine the effects of different sowing media on emergence and growth of moringa (*Moringa oleifera* Lam) seedlings in the nursery. The moringa seeds used for the study were collected from different parts of Nigeria; Nsukka (Eastern Nigeria), Ibadan (Western Nigeria) and Jos (Northern Nigeria). The sowing media were weathered sawdust (100%), top soil (100%), weathered sawdust plus cured poultry manure in the ratio of 2:1(volume by volume; v/v), and top soil plus cured poultry manure plus river sand in the ratio of 3:2:1(v/v/v). Perforated black polythene bags were used as potting media. The experiment was a 3 x 4 factorial trial in completely randomized design with three replications. The seeds sown in the 100% topsoil took average of 8.2

Corresponding author: E-mail: uchemay@yahoo.com;

and 9.0 days to have first and 50% seedling emergence. Weathered sawdust medium gave the highest mean percentage seedling emergence (84%) followed by the sawdust plus cured poultry manure (82%). The topsoil (control), 100% sawdust and weathered sawdust plus cured poultry manure had similar coefficient velocity of seedling emergence (11%). The topsoil medium (control) had the highest values of plant height and stem girth while sawdust plus poultry manure gave the highest number of leaves in the seedlings. There were no significant differences (P = .05) in the morphological traits of the seedlings in the different sowing media. The three accessions of Moringa oleifera also had no significant differences (P = .05) in their morphological growth. Sawdust and sawdust plus poultry manure encouraged early seedling emergence in moringa and sustained their growth in the nursery for four weeks. The media can, therefore, be used as good substitutes for topsoil in nursery establishment of crops with short nursery lives.

Keywords: Moringa; morphological growth; seedling emergence; sowing media.

1. INTRODUCTION

Moringa plants are easily established either by cuttings or seeds. Seeds are either sown directly in the field at the onset of the rainy season or planted in nurseries. Nursery operations involve raising seedlings in different media. Nursery potting media influence quality of seedlings produced thereof [1,2] which subsequently influences their establishment and productivity in the field [3]. The traditional nursery potting medium in Nigeria is topsoil dug from farmland and amended with poultry manure. Digging agricultural soils not only renders the land unproductive for cropping, but also makes it prone to erosion and other forms of degradation [4]. The quality of seedlings obtained is influenced by the composition of media used [5,6]. Ekwu and Mbah [6] reported on the relative importance of soiless media for growing potted ornamental plants in Nigeria. Baiyeri [7] evaluated three soiless media for weaning banana plantlets and showed that most of the genotypes produced vigorous seedlings when grown in rice hull (RH) composted with poultry manure (PM). Percentage germination, seedling emergence and growth in different sowing media are affected by the physical and chemical compositions of the media. Use of coarse materials in a sowing medium ensured greater aeration and drainage of the medium and also enhanced germination and seedling emergence [8].

Moringa is a fast-growing, deep-rooted dicotyledonous plant with tuberous taproot system. It is drought-tolerant and can thrive well in poor soils with little or no fertilization. There are 13 species of moringa which include Moringa oleifera Lam, M. arborea Verdc, M. borziana Mattei, M. concanensis Nimmo, M. drouhardii Jum, etc. Moringa oleifera is the most widely cultivated because of its high nutritional, medicinal, agricultural, domestic and environmental purposes. Moringa oleifera plants can be irrigated during the dry season but should not be water-logged to avoid root rot. The trees can grow up to a height of 15 m. The leaves are tri-pinnate, usually 25-60 cm long. The plants flower about three times a year and produce average of 17 tonnes of seeds/ha/annum in Nigeria [9]. The flowers are pollinated by bees. The seeds have very short dormancy period and can germinate promptly immediately after harvest. The seeds germinate within five to ten days of planting and the seedlings have short nursery life span (four to eight weeks) because of the fast growing nature of the plants [10]. The seeds lose their viability within one year of storage. Moringa can also be propagated using mature stem cuttings. The cuttings produce roots within two weeks of planting without any rooting hormone. The leaves are good sources of protein, minerals, vitamins, beta-carotene, amino acids and various phenolic compounds [11]. Moringa is widely cultivated in Nigeria. It is environmental-friendly because of its ability to absorb a lot of the atmospheric carbondioxide for its all-year-round leaf production (evergreen plant). It is a native of India but widely grown in parts of Afghanistan, Israel, Iran, Nepal, Bangladesh, China, Taiwan, Sri Lanka, Malaysia, Philippines, Thailand, Vietnam, Indonesia and Papua New Guinea [12]. Moringa plant does not perform well in intercrop with maize and cassava but can be used as alley crop in arable crop farming.

Some of the soiless materials used as sowing media, such as sawdust and rice husks, provide conducive environment for seed germination and seedling emergence as well as subsequent early seedling growth when mixed with manure [12]. High percentage seed loss has been observed in direct planting of moringa seeds in the field as many of them are destroyed by termites and rodents [12]. There is, therefore, the need to plant moringa in the nursery before transplanting them to the field to reduce losses and ensure that healthy and vigorous seedlings are transplanted into the field. The objective of this study was to determine the effects of different sowing media on emergence and early growth of moringa seedlings in the nursery.

2. MATERIALS AND METHODS

2.1 The Experimental Site

The experiment was carried out in the Teaching and Research Farm of the Department of Crop Science, University of Nigeria, Nsukka. Nsukka, is located on latitude 06° 52'North, longitude 07° 24'East and altitude 447.26 meters above sea level (m.a.s.l). Rainfall distribution pattern in this region is bimodal with peaks in July and September and a short dry spell around mid-August. The soil of the experimental site is a reddish-brown clay loamy Ultisol (Oxicpaleustult) belonging to the Nsukka series. It consists of 78% sand, 14% clay and 8% silt. Nsukka has derived savanna vegetation with mean annual minimum and maximum temperatures of 25 and 32°C respectively [13].

2.2 The Experimental Materials Used

The moringa seeds used for the study were collected from different parts of Nigeria; Nsukka (Eastern Nigeria), Ibadan (Western Nigeria) and Jos (Northern Nigeria) and they constituted the accessions. The sowing media were 100% weathered sawdust (SD), 100% top soil (TS), weathered sawdust (SD) plus cured poultry manure (PM) in the ratio of 2:1(volume by volume; v/v), and top soil plus cured poultry manure plus river sand (RS) in the ratio of 3:2:1(v/v/v). Perforated black polythene bags were used for potting the media.

2.3 Experimental Design and Cultural Practices in the Nursery

The experiment was a 3 x 4 factorial trial in completely randomized design with three replications. Seeds were collected from dry pods of the different moringa accessions and sown in the potted media at two centimeters depth. The seeds were unhulled (seed coat not removed) and not treated before planting. Ten seeds were planted per polythene bag. The nursery was not shaded but fenced round with one millimeter-mesh stainless wire gauze against rodents. The experiment was carried out during the dry

season, so the plants were not rain-fed. The media were watered routinely every other day at the rate of 300 ml/pot. Weeding was done by hand rouging. There was no fertilizer or pesticide application throughout the nursery period.

2.4 Seedling Emergence and Morphological Data Collection

Seedling emergence was recorded daily when the first foliage leaf appeared. Percentage seedling emergence was determined using the formula:

$$E = \frac{Number\ of\ emerged\ seedlings}{total\ number\ of\ seeds\ planted} x \frac{100}{1}$$

Where %E = percentage seedling emergence Coefficient velocity of seedling emergence (CVE) was calculated using Kotoski [14] formula stated as follows:

$$CVE = \frac{1}{N_1T_1 + N_1T_2 + \ldots N_xT_x} X \frac{100}{1}$$

Where N = Number of seeds emerging within the consecutive intervals of time. T = time between the beginning and end of a particular interval of measurement.

Days to first and 50% seedling emergence were recorded as the number of days from date of sowing to first and 50% seedling appearance.

The morphological growth parameters measured were plant height (cm), stem girth (cm) and number of leaves. The plant height was measured from the root/shoot junction to the shoot tip using a meter rule. The stem girth was determined by multiplying the stem diameter by 22/7 (π).

Stem Girth = Stem Diameter x 22/7 (π)

Number of the leaves produced was obtained by counting. All the morphological parameters were measured forth nightly for four weeks after planting.

2.5 Soil and Data Analyses

Physical and chemical properties of the sowing media were determined through laboratory analyses before sowing the seeds. Data collected were subjected to analysis of variance (ANOVA) according to Obi [15]. Table 1 shows the ANOVA Table used. The significant means were separated using Fisher's least significant difference (FLSD) at 5% probability.

Table 1. Analysis of variance (ANOVA) table used for data analysis

Sources of variation	General d.f	Specific d.f
Block	r - 1	3 − 1 = 2
Accessions (a)	a - 1	3 − 1 = 2
Media (m)	m - 1	4 − 1 = 3
a x m	(a − 1)(m - 1)	(3 -1)(4 - 1) = 6
Error	(am - 1)(r - 1)	(12 - 1)(3 - 1) = 22
Total	(amr - 1)	(3 x 4x 3) − 1 = 35

3. RESULTS AND DISCUSSION

3.1 Physical and Chemical Properties of the Sowing Media

The weathered sawdust plus poultry manure had the highest total porosity (71%) and the least bulk density (0.7 gm/cm^3). The 100% topsoil had the highest bulk density (1.6 g/cm^3) and least porosity (41%). The topsoil plus cured poultry manure plus river sand had the highest water holding capacity (35%) and the 100% weathered sawdust the least (28%). There were significant differences (P = .05) in the porosity of the different media (Table 2). The 100% sawdust and sawdust plus poultry manure had the highest organic carbon and organic matter contents (20% for organic carbon and 34 for organic matter each). The sawdust plus poultry manure had the highest values of total nitrogen (1.1%), potassium (0.5%), phosphorus (0.21) and pH in water (7.2) while the 100% topsoil had the least values of all the chemical properties except the total nitrogen content. There were significant differences (P = .05) in the chemical compositions of the sowing media except the pH values (both in water and KCl) as shown in Table 2.

3.2 Effects of Sowing Media and *Moringa oleifera* Accessions on Seedling Emergence

The seeds sown in the 100% topsoil took average of 8.2 and 9.0 days to have first and 50% seedling emergence while the seedlings in the topsoil plus poultry manure plus river sand emerged within five days (Table 3). Sawdust medium gave the highest mean percentage seedling emergence (84%) followed by sawdust plus poultry manure (82%). The topsoil (control), 100% sawdust and sawdust plus poultry manure had siimilar coefficient velocity of seedling emergence (11%). The topsoil plus poultry manure plus river sand gave the least values of both total percentage (16%) and coefficient velocity of seedling emergence (2.9%). There were significant differences (P = .05) in the total percentage and coefficient velocity of seedling emergence in the different media (Table 3). Seedlings of Nsukka accession emerged most promptly with the least average number of days to first and 50% emergence (6.2 and 6.8 days respectively). Jos accession had the highest percentage and coefficient velocity of emergence followed by Nsukka and Ibadan accessions respectively. There were significant differences (P = .05) in the total percentage and coefficient velocity of seedling emergence among the different accessions (Table 4). There were no significant accession by media interaction effects on seedling emergence, total percentage and coefficient velocity of emergence (Table 5).

3.3 Effects of Sowing Media and *Moringa oleifera* Accessions on Morphological Growth Traits

After four weeks of planting, the seedlings grown in 100% topsoil had the highest values of plant height and stem girth followed by those grown in 100% sawdust, sawdust plus poultry manure and topsoil plus poultry manure plus river sand in that order. Sawdust plus poultry manure gave the highest number of leaves in the seedlings while the mixture of topsoil, poultry manure and river sand gave the least. There were no significant differences (P = .05) in the morphological traits of the seedlings in the different sowing media (Table 6). The Nsukka accession gave the highest plant height values in the second and fourth week of planting while the Jos accession had the highest stem girth values at both periods of growth. The Ibadan accession had the least values of all the morphological growth traits at both periods. However, there were no significant differences (P = .05) in the growth traits of the seedlings of the three accessions all through the period of observation (Table 7). The accession by media interaction effects on the morphological growth traits were also not significant (Table 8).

Table 2. Physical and chemical properties of media used for raising *Moringa oleifera* seedlings in the nursery at University of Nigeria, Nsukka

Properties	Potting media			
	100% SD	2:1SD plus PM	100% (TS)	3:2:1TS plus PM Plus RS
Physical				
Total porosity (%)	70	71	41	47
Water holding capacity (%)	28	31	31	35
Bulk density (g/cm^3)	0.8	0.7	1.6	1.4
Chemical				
Organic carbon (%)	20	20	1.2	2.2
Organic matter (%)	35	35	2.0	3.7
Total nitrogen (%)	0.14	1.121	0.084	0.07
Potassium (%)	0.239	0.523	0.08	0.08
Phosphorus (%)	0.09	0.21	0.0029	0.0117
pH (H_2O)	6.7	7.2	6.2	6.8
pH (Kcl)	5.8	6.5	5.0	6.6

100% SD = Sawdust alone, SD plus PM = Sawdust plus poultry manure, TS plus PM plus RS = Topsoil plus poultry manure plus River sand, 100% TS = Topsoil alone

Table 3. Effects of media on the seedling emergence of *Moringa oleifera* plant in the nursery at University of Nigeria, Nsukka

Media	DE_1	DE_{50}	PE(%)	CVE(%)
TS	8.2	9.0	73	11
SD	8.0	8.4	84	11
SD plus PM	7.8	8.2	82	11
TS plus RS plus PM	5.1	5.1	8.9	4.2
LSD$_{(.05)}$	1.2	1.3	16	2.9

DE_1 = Days to first seedling emergence, DE_{50} = Days to 50% seedling emergence, PE = Percentage seedling emergence, CVE = Coefficient velocity of emergence, TS = Top soil, SD = Sawdust, PM = Poultry manure, RS = River sand, ns = Not significant, LSD$_{(.05)}$ = Least significant difference at 5% probability level

Table 4. Effects of accessions of *Moringa oleifera* on seedling emergence in the nursery at University of Nigeria, Nsukka

Accession	DE_1	DE_{50}	PE(%)	CVE(%)
Nsukka	6.2	6.8	68	10
Jos	7.9	8.2	70	10
Ibadan	7.8	8.2	48	8.3
LSD$_{(.05)}$	.2	.3	14	2.5

DE_1 = Days to first seedling emergence, DE_{50} = Days to 50% seedling emergence, PE = Percentage seedling emergence, CVE = Coefficient velocity of emergence, ns = Not significant, LSD$_{(.05)}$ = Least significance different at 5% probability level

Lower bulk densities and higher porosity of the 100% sawdust and the sawdust plus poultry manure (soiless) media might have aided prompt and higher coefficient velocity of emergence of the seedlings. Though the seedlings emerged most promptly in the topsoil plus poultry manure plus river sand medium, least percentage

emergence and coefficient velocity of emergence were obtained in the medium. This was probably the reason for the poor morphological (vegetative) growth of the seedlings in the medium. The seedlings had relatively higher percentage and coefficient velocity of emergence in the 100% sawdust and sawdust plus poultry manure compared to the other media. The two soiless media also compared favorably with the topsoil medium (control) in promoting early nursery growth of the seedlings. All these attributes qualify the sawdust and sawdust plus poultry as good growth media for crop seedlings with short nursery lives. The sawdust and sawdust plus poultry manure media compared favorably with the control in terms of the plant height, stem girth and number of leaves probably because they contained reasonable amounts of organic matter, nitrogen, phosphorus and potassium. Ekwu and Mbah [13] reported the relative importance of soiless media for growing potted ornamental plants in Nigeria. In Nigeria, sawdust is regarded as waste from carpenters' workshops and is often times burnt. Use of sawdust as nursery potting medium is a score in organic agriculture as a means of converting organic waste to useful agricultural material. It also reduces air pollution caused by smoke and greenhouse gases released in the process of burning, which deplete the ozone layer and result to global warming. The non-accessional differences obtained in the emergence and morphological growth of the seedlings suggest relative homogeneity in the genetic compositions of the seeds.

Table 5. Effects of accession by media interaction on moringa seedling emergence, total percentage and coefficient velocity of emergence at University of Nigeria, Nsukka

Accession	Media	DE_1	DE_{50}	PE (%)	CVE (%)
Nsukka	TS	7.7	8.7	80	11
	SD	7.3	8.0	100	12
	SD + PM	7.0	7.7	87	12
	TS + RS + PM	2.7	2.7	6.7	4.2
	Mean	6.2	6.2	68	10
Jos	TS	8.0	8.7	93	11
	SD	8.0	8.3	87	12
	SD + PM	7.7	7.7	87	12
	TS + RS + PM	8.0	8.0	13	5.9
	Mean	7.9	8.2	70	10
Ibadan	TS	9.0	9.7	47	10
	SD	8.7	9.0	67	11
	SD + PM	8.7	9.3	73	9.8
	TS + RS + PM	4.7	4.7	6.7	2.4
	Mean	7.8	8.2	48	8.3
$FLSD_{.05}$ for two media (M) means		1.2	1.3	16	2.9
$FLSD_{.05}$ for two accessions (A) means		.2	.3	14	2.5
$FLSD_{.05}$ for two interaction (M x A) means		1.3	1.3	28	5.1

Table 6. Effects of media on morphological growth traits of _Moringa oleifera_ at two and four weeks after planting in the nursery at University of Nigeria, Nsukka

Media	2 WAP			4 WAP		
	PHT(cm)	SG(cm)	NL	PHT(cm)	SG(cm)	NL
TS	7.9	0.71	3.7	14	0.95	6.0
SD	6.8	0.69	3.8	9.1	0.8	6.0
SD plus PM	6.1	0.60	3.9	10	0.81	6.4
TS plus RS plus PM	4.0	0.31	1.8	12	0.79	3.7
$LSD_{(.05)}$	1.1	.23	1.4	1.4	.54	2.8

TS = Topsoil, SD = Sawdust, SD plus PM = Sawdust plus poultry manure, TS plus RS plus PM = Topsoil plus River sand plus poultry manure, PHT = plant height, SG = stem girth, NL = Number of leaves, LSD(.05) = least significant difference at 5% probability level, ns = Not significant

Table 7. Effects of accessions on morphological growth traits of _Moringa oleifera_ plants at two and four weeks after planting in the nursery at University of Nigeria, Nsukka

Accession	2 WAP			4 WAP		
	PHT(cm)	SG(cm)	NL	PHT(cm)	SG(cm)	NL
Nsukka	6.9	0.61	3.7	13	0.93	6.0
Jos	6.3	0.62	3.7	12	0.97	6.4
Ibadan	5.0	0.51	2.5	8.2	0.67	5.3
$LSD_{(.05)}$	1.4	.12	.58	2.3	.26	1.3

PHT = Plant height, SG = Stem girth, NL = Number of leaves, LSD(.05) = Least significant difference at 5% probability level, ns = not significant

Table 8. Accession by media interaction effects on morphological growth   traits of *Moringa oleifera* at two and four weeks after planting

Accession			2 WAP			4WAP	
	Media	PHT	SG	NL	PHT	SG	NL
Nsukka	TS	8.3	0.72	4.0	15	1.03	8.7
	SD	7.9	0.71	4.0	11	0.79	5.7
	SD + PM	7.9	0.76	5.3	13	1.03	7.3
	TS + RS + PM	3.4	0.23	1.3	14	0.89	3.7
	Mean	4.2	0.61	3.7	11	0.94	6.3
Jos	TS	7.1	0.71	4.0	14	1.0	7.3
	SD	6.0	0.69	4.0	8.3	0.81	5.3
	SD + PM	6.7	0.62	4.0	12	0.85	7.7
	TS + RS + PM	5.4	0.47	2.7	15	1.2	5.3
	Mean	6.3	0.63	7.7	12	0.98	6.4
Ibadan	TS	6.7	0.70	3.0	14	0.81	8.0
	SD	6.5	0.68	3.3	8.4	0.82	7.0
	SD + PM	3.7	0.42	3.3	6.5	0.55	4.3
	TS + RS + PM	3.1	0.23	1.3	4.2	0.26	2.0
	Mean	5.0	0.51	2.8	8.2	0.61	5.3
$FLSD_{.05}$ for two media (M) means		1.1	0.23	1.4	1.4	.54	2.8
$FLSD_{.05}$ for two accessions (A) means		1.4	.12	.58	2.3	.26	1.3
$FLSD_{.05}$ for two interaction (M x A) means		2.2	.26	.89	3.2	.36	2.4

4. CONCLUSION

Weathered sawdust and sawdust plus poultry manure gave the highest total percentage emergence and compared favorably with the topsoil medium (control) in terms of coefficient velocity of seedling emergence, plant height, stem girth and number of leaves. They can, therefore, be used as suitable alternatives to topsoil for raising crop seedlings with short nursery life spans.

COMPETING INTERESTS

Authors have declared that no competing interests exist.

REFERENCES

1. Agbo CU, Omaliko CM. Initiation and Growth of shoots of *Gongronema latifolia* (Benth) stem cuttings in different rooting media. Afr. J. Biotechnol. 2006;5(5):25-28.
2. Sakin UO, Ercisli S, Anapali O, Esitken A. Effect of Pumice amendment on physical soil properties and strawberry plant growth. J. Central Europe. Agric. 2005;6(3):361-266.
3. Becker K, Foild N, Makkar HPS. The Potential of *Moringa oleifera* for agricultural and industrial uses, In: The miracle tree -

the multiple attributes of *Moringa* (d. Lowel J. Fugue). CTA. USA; 2001.
4. Baiyeri KP. Evaluation of nursery media for seedling emergence and early seedling growth of two tropical tree species, Monr J. Agric. Res. 2003;4(1):60-65.
5. Corti C, Crippa L, Genevini PL, Centemero M. Compost use in plant nurseries, hydrological and physiochemical characteristics, Compost Sci. Utilization. 1998;6:35-45.
6. Wilson SB, Stoffella PJ, Graetz DA. Use of Compost as a media amendment for containerized production of two subtropical perennials. J. Environ Hort. 2001;19(1):37-42.
7. Baiyeri KP. Response of Musa species to micro-propagation; 11: The effects of genotypes, initiation and weaning media on sucker growth and quality in the nursery, Afri. J Biotechnol. 2005;4(3):229-234.
8. Ndubuaku UM, Oyekanmi EO. Preliminary effect of sawdust/topsoil mixtures on germination, root volume and other growth parameters of F_3 Amazon cocoa seedlings in the nursery. Nig J Sci. 2000;34(4)389-394.
9. Ndubuaku UM, Ndubuaku TCN, Ndubuaku NE. Yield characteristics of *Moringa oleifera* across different ecologies in Nigeria as an index of its adaptation to

climate change. Sustainable Agric. Res. 2014;3(1):95-100.

10. Fuglie LJ. The Miracle Tree: *Moringa oleifera*; National nutrition for the tropics. church world service. Dakar, revised in 2001 and published as the Miracle Tree. The Multiple Attributes of Moringa; 1999. Available:http://www/echotech.org/booksto re/advanced

11. Fuglie LJ. The Moringa tree: A local solution to malnutrition Published in Dakar, Senegal; 2005.

12. Papillo J. *Moringa oleifera*: the multipurpose wonder-tree. Michigan Technological University, Michigan, USA; 2007.

Abailable:http://peacecorps.mtu.edu/resour ces/studentprojects/moringa.htm

13. Ekwu LG, Mbah BN. Effects of varying levels of nitrogen fertilizer and some potting media on the growth and flowering response of Marigold (*Targetes erecta* L), Nig. J. Hort. Sci. 2001;5:104-109.

14. Kotoski F. Temperature relations to germination of vegetables seed proc. Amer. Soc Hot. Sci. 1978;23:176-184.

15. Obi IU. Statistical methods of detecting differences between treatment means and research methodology issues in laboratory and field experiments. 2nd ed; 2002.

16

Profit Efficiency of Cocoyam Production in Kaduna State, Nigeria

S. Abdulrahman[1*], O. Yusuf[1] and A. D. Suleiman[2]

[1]Department of Agricultural Economics and Rural sociology, Faculty of Agriculture,
Ahmadu Bello University, Zaria, Kaduna State, Nigeria.
[2]Department of Agricultural Education, Federal College of Education Okene, Kogi State, Nigeria.

Authors' contributions

This work was carried out in collaboration between all authors. Author SA designed the study, wrote the protocol and wrote the first draft of the manuscript. Author OY reviewed the experimental design and all drafts of the manuscript and managed the analyses of the study. Author ADS identified the plants and performed the statistical analysis. All authors read and approved the final manuscript.

Editor(s):
(1) Rusu Teodor, Department of Technical and Soil Sciences, University of Agricultural Sciences and Veterinary Medicine Cluj-Napoca, Romania.
Reviewers:
(1) Anonymous, USA.
(2) Anonymous, Turkey.

ABSTRACT

Aims: Aims of the study were to describe socio-economic characteristics of cocoyam farmers, profitability of cocoyam production and determine the profit efficiency of cocoyam producers in Kaduna state.
Study Design: Primary data were collected from cocoyam producers through the use of structured questionnaires.
Place and Duration of Study: This study was carried out in three local government area in Kaduna state, Nigeria between August and November 2014 cropping season.
Methodology: Multistage purposive and random sampling techniques were employed for data collection.
Results: The study showed that 34% of the respondents fall within the age range of 30 and 39years. The majority of the farmers (50%) had no formal education. The household size ranged from 6-10 persons, whereas (73%) were not members of cooperative society. Results indicated that except for cost of fertilizer, all other factors were significant (P < 0.01 and P < 0.1). The mean profit efficiency is 66% while the range is 3-99%

Corresponding author: E-mail: arsan4u@gmail.com;

Conclusion: The findings of the study revealed that none of the sampled cocoyam farms reached the frontier threshold. Also, amount of credit received and farming experience was the socio-economic variable responsible for the variation in profit efficiency of the cocoyam producers. It was therefore recommended that timely and adequate supply of seed should be made available to farmers at affordable price in order to increase profit from production of cocoyam.

Keywords: Profit efficiency; cocoyam; stochastic production frontier; Kaduna state.

1. INTRODUCTION

Nigeria's domestic economy is partly determined by agriculture which accounted for 40.9% of the Gross Domestic Product (GDP) in 2010 [1]. Agriculture has been an important sector in the Nigerian economy in the past decades and is still a major sector despite the oil boom. Basically it provides employment opportunities for the teeming population, eradicates poverty and contributes to the growth of the economy. Despite these however, the sector is thus characterized by low yields, low level of inputs and limited areas under cultivation [2]. Nigeria is an agrarian economy with 70% of its people dependent on agriculture [3]. The Government of Nigeria has been trying to achieve food security at both house hold and national level through its mechanized approach.

Root and tuber crops which are among the most important groups of staple foods in many tropical African countries [4] consistute the largest source of calories for the Nigeria population [5]. Cassava (*Manihot esculenta*) is the most important of these crops in terms of total production, followed by yam (*Dioscorea* spp.), cocoyam (*Colocasia* spp. and *Xanthosoma* spp.) and sweet potato (Ipomoea batatas) [5].

Cocoyam (*Colocasia esculenta* and *Xanthosoma mafafa* (L) Okeke) are important carbohydrate staple food particularly in the southern and middle belt areas of Nigeria [6]. Nutritionally cocoyam is superior to cassava and yam in the possession of higher protein, mineral and vitamin contents in addition to having more digestible starch [7,8]. Cocoyam which ranks third in importance and extent of production after yam and cassava is of major economic value in Nigeria [9]. Edible cocoyam cultivated in the country is essentially species of *Colocasia* (taro) [10] and *Xanthosoma* (tannia). The average production figure for Nigeria is 5,068,000 mt which accounts for about 37% of total world output of cocoyam [11]. Small scale farmers, especially women who operate within the

subsistence economy grow most of the cocoyam in Nigeria.

It is highly recommended for diabetic patients; the aged, children with allergy and for other persons with intestinal disorders [12]. According to [13], boiled cocoyam corms and cormels are peeled, cut up, dried and stored or milled into flour. The flour can be used for soups, biscuits, bread and puddings for beverages. The peels can also be utilized as feed for ruminants.

Despite the importance of cocoyam, more research attention has been given to cassava and yam [14,15]. [16], Observed that research on cocoyam has trailed behind cassava and yam as root crops in Nigeria and other countries. [17] noted that the totality of published scientific work on cocoyam is insignificant when compared with those of rice, maize, yam and cassava. However, [16] asserted that it was only in the last decade that policy makers and national agricultural research systems began to show systematic interest in the crop because of concern over biodiversity. There is a declining trend in cocoyam production as well as a shortage of its supply in domestic markets as a result of a number of technical, socio-economic and institutional constraints, which need to be addressed. Cocoyam farmers are generally found on a small scale and its production has been undermined.

Arising from the forgoing, there is need to have a look into the production of cocoyam, one of the major roots and tuber crops in Nigeria which is fast becoming an extinction crop. This is due to the general believed that most families no longer consume it because it is not readily available for consumption even during its season, as a result of reduction in its production level. This study therefore measured the technical efficiency of cocoyam producers in Kaduna state.

Production of cocoyam has not been given priority attention in many countries probably because of its inability to earn foreign exchange and its unacceptability by the high income countries for both consumption and other purposes [18]. Most of what is produced is consumed locally [19]. The production is labour intensive with most operations carried out manually at the traditional level. There is a dearth of information on the economics of cocoyam production in Nigeria.

1.1 Concept of Efficiency Measurement Using Frontier Profit Function

[20] in his pioneering study defined efficiency as the ability to produce a given level of output at lowest cost. Efficiency can be analyzed by its two components: technical and allocative efficiency. Technical efficiency is defined as the degree to which a farmer produces the maximum feasible output from a given bundle of inputs (an output oriented measure), or uses the minimum feasible of inputs to produce a given level of output (an input oriented measure). On the other hand, allocative efficiency relates to the degree to which a farmer utilizes inputs in optimal proportions, given the observed input prices [21]. These components have been measured by the use of frontier production function which can be deterministic or stochastic. Deterministic frontier production function explains that all deviations from the frontier are attributed to inefficiency where as in stochastic frontier production function it is possible to discriminate between random errors and differences in efficiency [21]. [22] argued that a production function approach to measure efficiency may not be appropriate when farmers face different prices and have different factor endowments [23]. Thus, this led to the application of stochastic profit function models to estimate farm specific efficiency directly [23,21,24,25,26].

According to [26] the profit function approach combines the concepts of technical and allocative efficiency in the profit relationship and any error in the production decision is assumed to be translated into lower profits or revenue for the producer. Profit efficiency is defined as the ability of a farm to achieve highest possible profit given the prices and levels of fixed factors of that farm and profit inefficiency is defined as loss of profit from not operating on the frontier [23]. It should be noted that [27] had extended the stochastic production frontier model by suggesting that the inefficiency effects can be expressed as a linear function of explanatory variables, reflecting farm-specific characteristics. The advantage of their model is that it allows estimation of the farm-specific efficiency scores and the factors explaining efficiency differentials

among farmers in a single stage estimation procedure. This study therefore, used [27] model by postulating a profit function, which is assumed to behave in a manner consistence with the stochastic frontier concept. The model was applied to cocoyam producers in Kaduna State, Nigeria.

The stochastic frontier profit function is defined as:

$$\pi_i = f(X_i; \delta) + \varepsilon_i \dots \dots \dots \dots \dots \qquad \dots \dots \dots (1)$$

Where π normalized profit of the ith farms is, X_i is a vector of inputs used by farm i, and ε_i is a "composed" error term. The error term ε_i is equal to $v_i - u_i$. The term v_i is a two-sided ($-\infty < v_i < \infty$) normally distributed random error ($v \sim N[0, \sigma_v^2]$) that represents the stochastic effects outside the farmers' control. The term u_i is a one-sided ($u_i \geq 0$) efficiency component that represents the technical inefficiency of farm. The distribution of the term u_i can be half-normal, exponential, or gamma [28,29] and half-normal distribution ($u \sim N[0, \sigma_u^2]$) is used in this study. The two components v_i and u_i are also assumed to be independent of each other.

Profit efficiency (PE) of an individual firm is defined in terms of the ratio of predicted actual profit to the predicted maximum profit for a best-practiced cocoyam farmer, conditioned on the level of price of output and inputs used by the firm. Profit inefficiency is therefore defined as the amount by which the level of profit for the firm is less than the frontier profit. This is shown in equation [2]

$$PE_i = \frac{\pi}{\pi^{max}} = \frac{f(X_i\delta)\exp(V_i - U_i)}{f(X_i\delta)\exp(V_i)} = \exp(-U_i) \quad (2)$$

2. MATERIALS AND METHODS

2.1 Study Area

This study was conducted in Kaduna state of Nigeria. Kaduna state lies in the north western part of the country's geopolitical zone, about 200km away from Abuja the federal capital. The state lies between latitudes 90°N and 12°N of the equator and between longitudes 6°E and 9°E of the prime meridian. Kaduna state shares boundaries with Katsina and Kano state to the north. Plateau to the north east, Nasarawa and Abuja to the south and Niger and Zamfara state to the west [30]. The state occupies an area of approximately 68,000 square kilometers or 7% of Nigeria's land mass. The state has 23 Local

Government Areas [31]. The mean annual rainfall shows a marked decrease from South to North (1,524 mm to 635 mm). Two distinct seasons occur in the state; the rainy season and the dry season. The relative humidity is constantly below 40% except in few wet months when it goes up to an average of 60%. The duration of dry season is 5-7 months which normally starts from October. The state is agrarian and well suited for the production of arable crops such as maize, yam, millet, and sorghum because of a favourable climatic condition. Livestock production is also practiced in the state. Rearing of goats, sheep, cattle and different classes of poultry as well as marketing of their products is practiced in the state. The people of the state live mostly in organised towns and cities [32]. A large variety of non-agricultural occupations also exit.

The total population of the state is 6.11 million [31]. Based on annual population growth rate of 3.2%, the projected population of the state was about 7.33 million people in 2012. Within the state there are a number of establishments ranging from companies, research institutes, higher institutions and colleges.

2.2 Sampling Procedure

Multistage sampling techniques were used to select respondents for this study. The first stage involved a purposive selection of the three (Giwa, Kudan and Ikara) local governments based on predominance of cocoyam production among the farmers. Secondly, 9 villages were purposively selected, Three (Giwa, Yakawada, Guga; Gimbawa, Kwasallawa, Malikanchi; and Musawa, Hunkuyi, Kudan) from each local government area based on their intensity of cocoyam production. Finally, a simple random sampling was employed in selecting farmers from each of the villages. Fifty percent (50%) of the sample frame (248) was used as the sample size. In all, 124 farmers were randomly selected for the study.

2.3 Data Collection and Analysis

Primary data were used for this study. These were collected with the aid of structured questionnaires. The information collected includes labour input, fertilizer input, seed, farm size and farmer's socio-economic characteristics such as age, household size, educational status, amount of credit received, number of extension contacts, years spent on the cooperative, and income.

2.4 Model Specification

Empirical model specification for the determinants of profit efficiency is as follows;

$$\ln\pi_i = \beta_0 + \beta_1 \ln X_{1i} + \beta_2 \ln X_{2i} + \beta_3 \ln X_{3i} + \beta_4 \ln X_{4i} + V_i - U_i$$

Where subscript i refer to the observation of ith farmers,

ln = Logarithm to base e,
π_i = Profit of the ith farmers (₦)
X_1 = average price of seed (₦)
X_2 = average price of fertilizer (₦)
X_3 = average price of labour (₦)
X_4 = average price of farm size (₦)

The inefficiency effects, V_i is a random error term assumed to be independently and identically distributed as N $(0, \sigma_v^2)$. U_i represents profit inefficiency and is identically and distributed as a truncated normal with truncations at zero of the normal distribution [33]. The U_i is defined as:

$$U_i = \delta_0 + \delta_1 \ln Z_{1i} + \delta_2 \ln Z_{2i} + \delta_3 \ln Z_{3i} + \delta_4 \ln Z_{4i} + \delta_5 \ln Z_{5i} + \delta_6 \ln Z_{6i}$$

Where:

U_i = Technical inefficiency of the ith farmer
Z_1 = Age of the farmer (years)
Z_2 = Years of education of the ith farmer
Z_3 = Household size of the ith farmer (Numbers of people)
Z_4 = Cooperative Association of the ith farmer (Years of participation)
Z_5 = Extension Contact of the ith farmer (Number of contacts)
Z_6 = Access to Credit by the ith farmer

3. RESULTS AND DISCUSSION

The socio-economic characteristics of the respondents are presented in Table 1. The study revealed that 34% of the respondents fall within the age range of 30 and 39 years. The mean age of the farmers was 40 years; this implies that the majority of the farmers were younger, who can contribute positively to agricultural production for the next two decades. This result is consistent with the findings of [34] who observed that youth constitute the majority of the cocoyam farmers, and younger farmers are more flexible to new ideas and risk; hence they are expected to adopt innovations more readily than older farmers. The majority of the farmers (50%) had no formal education. This indicates that the farmers' educational level is low. According to [35], education has a positive and significant impact on farmers' efficiency in production. The literacy level greatly influences the decision making and adoption of innovation by farmers, which may bring about increase in production of the crop. The educational level of farmers does not only increase his productivity but also increase his ability to understand and evaluate new techniques. The majority of the farmers (30%) had household size with 6-10 members. The average household size was 13 persons implying that there is appreciable source of family labour supply to accomplish various farm operations. According to the report of [36], there is a positive and significant relationship between household size and farmers' efficiency in production. However, the absolute number of people in a certain family cannot be used to justify the potential for productive farm work. This is because it can be affected by some important factors namely; age, sex and health status. This shows that a reasonable number of the respondents have a large household size. Higher household size provides enough persons for family labour and less money will be needed to pay for hired labour. About (73%) of cocoyam farmers do not participate in any cooperative association. According to them, their non-membership is due to being small scale and unawareness of any association while 27% participated with average of 2.4 times per year. The effect of this result is that most of the cocoyam farmers in the study area do not enjoy the assumed benefits accrued to co-operative societies through pooling of resources together for a better expansion, efficiency and effective management of resources and for profit maximization. [37] stated that membership of cooperative societies have advantages of accessibility to micro-credit, input subsidy and also as avenue in cross breeding ideas and information. (85%) of cocoyam farmers in the study area have no access to extension service while (15%) have access to extension service with average of 0.4/ year. This could be attributed to low extension agent-farmers' ratio in the study area.

3.1 Profitability of Cocoyam Production in the Study Area

The result in Table 2 revealed that cocoyam seed used by the farmers in the study area were mainly unimproved seeds taken from the last

harvest. The quantity of cocoyam set (seed) was 1068.2 kg/ha with an average market price of ₦271 per kg was used and this constitutes 62.2% of the total cost of production. The quantity of fertilizer was 490.47kg/ha with an average market price of ₦100 per kg was used and this constitutes 28.5% of the total cost of production.

Labour costs consist of cost of land preparation, planting, fertilizer application, weeding, replacement and harvesting. The family labour was computed on the basis of opportunity cost in man-day. The wage rate varied according to farm operation to be performed. An average wage rate of ₦400 per man-day was used, giving the average labour cost per hectare to be ₦9780 while the total cost of fixed inputs (cost of renting land and depreciation of tools) incurred on cocoyam production was ₦6113 and this constitute 3.6% of the total fixed cost.

The result in Table 2 revealed that the total revenue (TR) was ₦290,076.7 while the total cost (TVC + TFC) was ₦171,760. The net farm income was therefore ₦118,316.7. the average rate of return on investment (return per naira invested) was 1.69, indicating that for every ₦1 invested in cocoyam production in Kaduna state, a profit of 69 kobo was made. Thus, it could be concluded that cocoyam production in the study area though on a small scale, was economically viable. This finding is similar to that of [38] who observed that cocoyam production is profitable by returning ₦1.80 to every ₦1.00 spent.

3.2 Profit Efficiency and its Determinants among the Cocoyam Farmers

The maximum likelihood estimates of the parameters of the stochastic profit frontier model are presented in Table 3. The estimated sigma squared (σ^2) was significantly different from zero at the 1 percent level; this indicates a good fit and correctness of the specified distributional assumptions of the composite error terms. This conforms to [39,21]. In addition, the estimated gamma parameter (γ) of 0.30 was significant at 1 percent level of significance (Table 3), indicating that about 30 percent of the variation in actual profit from maximum profit (profit frontier) among cocoyam farms was due mainly to differences in farmers' practices rather than random variability.

Table 1. Socio-economic characteristics of cocoyam farmers

Variable	Frequency (N = 124)	Percentage
Age (years)		
20-29	32	25.8
30-39	42	33.8
40-49	17	13.7
50-59	20	16.0
60 above	13	10.4
Mean	40	
Educational status		
No formal education	62	50.0
Primary education	11	8.9
Secondary education	34	27.4
Tertiary education	17	13.7
Household size		
1-5	28	22.5
6-10	37	29.8
11-15	23	18.5
16-20	19	15.3
21 above	17	13.6
Mean	13	
Membership of cooperative society		
Non members	90	72.6
1-5	21	16.9
6-10	4	3.2
11-15	4	3.2
16 above	5	4.0
Mean	2	
Extension contact		
No contact	105	84.7
1-3	16	12.8
4-6	3	2.4
Mean	0.4	
Access to credits		
Personal savings	116	93.5
Borrowing	8	6.5

N = Number of respondents

Table 2. Average cost and return per hectare of cocoyam production

Variables	Values/ha (₦)	% Contribution
Total Revenue	290076.7	
Total cost (TVC + TFC)		
a. seed (kg)	106,820	62.2
b. fertilizer (kg)	49,047	28.5
c. labour (man-days)	9780.00	5.7
Total variable cost (a + b + c)	165,647	
a. Cost of renting land	4813	2.8
b. Depreciation of tools (hoe and cutlass)	1300	0.8
Total fixed cost= (a + b)	6,113	3.6
Total cost =(165,647 + 6,113)	171,760	
Net Farm Income= (NFI)= (TR - TC)	118,316.7	
Return per Naira Invested (TR/TC)	1.69	100

Only the coefficients of the cost seeds and cost of fertilizer were found to be positive while the cost of labour and cost of farm land were negative. Demonstrating that cost of seed and cost of fertilizer had positive effect on the profit efficiency of cocoyam farming in Kaduna state while the cost of labour and cost of farm land had negative effect on profit efficiency .

This implied that a unit increase in the prices of inputs with positive coefficient will lead to increase in the profit efficiency of cocoyam and vice versa. However, the coefficient for cost of seed with positive coefficient of 0.58 was statistically significant at 1 percent level of significance and this appears to be the most important variable determining profit efficiency. This implies that for a 1 percent increase the use of seed, the profit obtainable from cocoyam production will increase by 52 percent.

The negative sign of labour may be due to high cost of a negative relationship do exist between family labour and hired labour among the resource-poor rural farmers because the consumption of additional hired labour is meant to supplement available family labour such that as the availability of family labour decreases, additional hired labour is consumed at the limit of the lean resources of the farmers. Due to the high cost of hired labour if additional hired labour must be consumed then additional cost must be incurred while the negative cost of fertilizer may perhaps be due to wrong use leading to too much application of fertilizer by the cocoyam farmers, therefore resulting in extra cost sustained by the farmers.

The negative sign of farm land at 10 percent level of significance shows that for a 10 percent increase in the cost of farm land, the profit obtainable from cocoyam production will decrease by 37 percent. This may be due to over utilization of resources as a resulting of additional cost incurred; hence increasing their farm size will decrease profit, other things being equal. This finding is at variance with [40], significant coefficient of farm size at 5 percent level of significance points to the fact that cassava farmers were operating at small scale level, hence increasing their farm size will improve profit.

The parameters estimates for determinants of profit efficiency were reported in the lower part of Table 3. However the analysis of inefficiency models shows that the signs and significance of the estimated coefficient in the inefficiency model have important implication on the profit efficiency of the farmer.

The results further showed that the profit inefficiency of the cocoyam farmers was positively influenced by age, household size, cooperative membership and extension contact while education, farming experience, and credit negatively influence profit inefficiency (Table 3). This result is in agreement with [41,23]. Thus, investments in rural education through effective extension delivery program and provision of credit will boost farmers' efficiency. The result of this study has clearly shown that opportunities exist in cocoyam production.

The result showed that there is a significant and positive relationship between age and profit at 10% level of probability. This implies that cocoyam farmers with more age exhibited significantly less profit than farmers with less age.

The result showed that there is a significant and negative relationship between experience and profit at 10% level of probability. This implies that cocoyam farmers with more years of experience exhibited significantly more profit than farmers with less years of experience. This could probably be explained by the fact farmers probably employ their experience over time as an opportunity to enhance more profit. This finding is consistent with [42].

The result in Table 3 showed that there is a significant and negative relationship between credit and profit at 10% level of probability. This implies that cocoyam farmers with access to credit exhibited significantly more profit than farmers with less credit. Credit is a very strong factor that is needed to acquire or develop any enterprise; its availability could determine the extent of production capacity. It also agrees with findings of [43] who noted that access to micro-credit could have prospect in improving the productivity of farmers and contributing to uplifting the livelihoods of disadvantaged rural farming communities. This finding also conform to the study of [44] supported this fact by reporting in his study that credit increases the net revenue obtained from fixed inputs, market conditions and individual characteristics, while credit constraint decreases the efficiency of farmers by limiting the adoption of high yielding varieties and the acquisition of information needed for increased productivity.

The result in Table 3 showed that there is a significant and positive relationship between extension contact and profit. This implies that cocoyam farmers with more extension contact exhibited significantly less profit than farmers with less extension contact.

3.3 Frequency Distribution of Profit Efficiency Estimates of Cocoyam Farmers

Table 4 presents the distribution of profit efficiency of cocoyam farmers. The profit efficiency score ranged between 0.13 and 0.98 with an average of 0.66. The average profit efficiency score of 0.66 implied that an average

cocoyam farmer in the study area could increase profits by 34% by improving technical and allocative efficiency in cocoyam production. This result conformed to the findings of [45,21] who reported mean profit efficiency levels of 0.77 for Bangladeshi rice farmers and 0.78 for Nigerian cowpea farmers respectively. This result indicates that about 56% of cocoyam farmers seemed to be skewed towards efficiency level of 61% and above, while the farmer with the best and least practice had a profit efficiency of 0.98 and 0.13 respectively. In spite of this, the results implied that a considerable amount of profit can be obtained by improving technical and allocative efficiency in cocoyam production in the area

Table 3. Maximum likelihood estimates results of frontier profit function of cocoyam production

Variables	Parameters	Coefficients	Std. error	T-value
Profit Function				
Constant	β_0	15.497	1.517	10.213***
In cost of seed	β_1	0.576	0.162	3.563***
In cost of Fertilizer	β_2	0.194	0.236	0.823
In cost of Labour	β_3	-0.276	0.108	-2.551***
In cost of farm land	β_4	-0.375	0.204	-1.831*
Inefficiency variable				
Constant	Z_0	0.082	0.451	0.181
Age	Z_1	0.019	0.010	1.911*
Educational status	Z_2	-0.548	0.291	-1.887*
Household size	Z_3	0.002	0.015	0.131
Farming experience	Z_4	-0.005	0.019	-0.286
Cooperative association	Z_5	0.025	0.039	0.664
Amount of credit borrowed	Z_6	-0.00009	0.00005	-1.789*
Extension contact	Z_7	0.00006	0.00003	1.881*
Diagnostic Statistic				
Sigma-square	(σ^2)	0.681	0.094	7.223***
Gamma	(γ)	0.301	0.097	3.105***
Log likelihood function	L/f	-142.820		
LR test	31.633			
Total number of observation	164			
Mean efficiency	0.66			

Asterisk indicate significance ***1%,**5%, *10%.

Table 4. Frequency distribution of profit efficiency estimates from the stochastic frontier model

Efficiency level	Frequency	Percentage
< 0.2	11	8.87
0.21-0.40	8	6.46
0.41-0.60	41	33.06
0.61-0.80	28	22.58
0.81-1.00	36	29.03
Total	124	100
Minimum	0.13	
Maximum	0.99	
Mean	0.66	

4. CONCLUSION

The paper estimates the farm level profit efficiency and its determinants using the stochastic parametric method of estimation. The findings of the study revealed that none of the sampled cocoyam farms reached the frontier threshold. Also, amount of credit received and farming experience was the socio-economic variable responsible for the variation in profit efficiency of the cocoyam producers.

5. RECOMMENDATIONS

The coefficient for cost of seed with positive coefficient of 0.58 was statistically significant at 1 percent level of significance and this appears to be the most important variable determining profit efficiency. This implied that a unit increase in the prices of seed will lead to increase in the profit of cocoyam; it was therefore recommended that timely and adequate supply of seed should be made available to farmers at affordable price in order to increase profit from production of cocoyam. Also, the level of profit efficiency of some farmers was very low due to improper management of resources; it is therefore recommended that farmers should be trained and advised on proper and efficient utilization of resources (seed, farm size and labour) in order to improve their profit efficiency.

ACKNOWLEDGEMENTS

I specially wish to express my deep appreciation and sincere gratitude to Prof. S.A. Sanni and Prof. Z. Abdulsalam for their invaluable assistance, close supervision, constructive criticisms, suggestions and pieces of advice that aided the completion of this research work.

COMPETING INTERESTS

Authors have declared that no competing interests exist.

REFERENCES

1. Central Bank of Nigeria. Annual Report. CBN, Abuja, Nigeria; 2011.
2. Izuchukwu O. Analysis of the Contribution of Agricultural Sector on the Nigerian Economic Development. World Review of Business Research. 2011;1(1):191- 200.
3. NBS Facts and Figures about Nigeria. National Bureau of Statistics .Abuja, Nigeria; 2007.
4. Osagie PI. Transfer of Root Crop Technology for Alleviation of Poverty; the Contribution of Shell, Nigeria. In; Akoroda MO, Ekanayake IJ (eds.). Proceedings of the 6th Triennial Symposium of the International Society for Tropical Root Crops. 1998;38- 41.
5. Olaniyan GO, Manyoung VM, Oyewole B. The Dynamics of the Root and Tuber Cropping Systems in the Middle belt of Nigeria. In: Akoroda, M. O. and Ngeve, J. M. (eds.). Proceedings of the 7th Triennial Symposium of the International Society for Tropical Root Crops (ISTRC). 2001;75- 81.
6. Asumugha GN, Mbanaso ENA. Cost Effectiveness of Farm gate Cocoyam Processing into Frizzles. In Agriculture, a basis for poverty eradication and conflict resolution. Proc. of the 36th Annual Conference of Agricultural Society of Nigeria (ASN), Federal University of Technology Owerri (FUTO), Imo state, Nigeria. 2002;94-97.
7. Parkinson S. The Contribution of Aroids in the Nutrition of People in the South Pacific. In: Chandra, S (ed.). *Edible Aroids*. Clarendon Press, Oxford, U.K; 1984.
8. Splitstoesser NE, Martin FW, Rhodes AM. The Nutritional Value of Stochastic Frontier Production and Cost Function Estimation. Department of Economics, University of New England, Armidale, Australia. The Global Food System: A Vision Statement of the Year 2020. International potato Centre: Lima, Peru; 1973.
9. Udealor A, Nwadukwe PO, Okoronya JA. Management of Crop Xanthosoma, Alocasia, Crystosperma and Amorpholosphallus. In Tropical Root Crops; 1996.
10. Howeler RH, Ezumah HC, Midmore DJ. Tillage Systems for root Indian Agriculture. American Economic Review. 1993;61(1): 94-109.
11. FAO. Food and Agricultural Organisation Database Results; 2011. Available:http://www.fao.org/docrep/014/i2 215e/i2215e.pdf
12. Plucknet DC. The Status and Future of Major Aroids (*Colocosia, Xanthosoma, Alocasia, Crystosperma* and *Amorpholophallus*). In Tropical Root Crops Tomorrow. Proceedings of International

Symposium on Tropical Root Crops. Hawaii. 1970;(1) :127-135.

13. Ene CI. A Comparative Study of Fadama and Non- Fadama Crop farmers in Osisioma-Ngwa L.G.A, Abia State, Nigeria. Journal of Sustainable Tropical Agriculture Research11; 1992.

14. International Institute of Tropical Agriculture. Sustainable Food Production in sub- Saharan Africa. IITA's Contributions. International Institute of Tropical Agriculture, Ibadan, Nigeria; 1992.

15. Tambe RE. The economics of cocoyam production by small holder farmers in Manya Division, South west province of Cameroun, M.Sc. Project Report Department of Agricultural Economics, University of Nigeria, Nsukka; 1995.

16. Skott GJ, Best R, Rosegrant M, Bokanga M. Root and Tubers in Some Tropical Root Crops. Proceedings of the Tropical Region of the American Society for Horticultural Sciences south pacific.17:290-294.In: Chandra, S (ed.). Edible Aroids. Clarendon Press, Oxford, U.K. 2000;9:215-224.

17. Ezedinma FO. Prospects of Cocoyam in the Food System and Economy of Nigeria. In: Arene, Ene LSO, Odurukwe SO, Ezeh NO. A (eds.). Proceedings of the1st National Workshop on Cocoyam. 1987; 28- 32.

18. Onyenweaku CE, Ezeh NOA. Trends in Production, Area and Productivity of Cocoyams inNigeria 1960/61 – 1981/84: In Cocoyams in Nigeria, Production, Processing and Utilization, NRCRI Umudike. 1987;94-100.

19. Mbanaso ENA, Enyinnaya AM. Cocoyam germplasm conservation. NRCRI Annual Report. 1989;60.

20. Farrell MJ. The Measurement of Productive Efficiency. Journal of the Royal Statistical Society. Series A (General). 1957;120(3):253-290.

21. Rahman S. Profit efficiency among Bangladeshi rice farmers. Food Policy. 2003;28:483-503.

22. Yotopolous PA, Lau LJ. A test for relative economic efficiency: Some further results. American Economic Review. 1973;63(1): 214-223.

23. Ali M, Flinn J. Profit efficiency among Basmati rice producers in Pakistan Punjab. American Journal of Agricultural Economics. 1989;71(2):303- 310.

24. Wang J, Cramer GL, Wailes EJ. A Shadow-Price Frontier Measurement of Profit Efficiency In Chinese Agriculture. American Journal of Agricultural Economics. 1996;78:146-156.

25. Ogundari K, Ojo SO, Ajibefun IA. Economies of scale and cost efficiency in small scale maize production: empirical evidence from Nigeria. Journal of social science. 2006;13(2):131-136.

26. Ali F, Parikh A, Shah MK. Measurement of profit efficiency using behavioral and stochastic frontier approaches. Journal of Applied Econometrics. 1994;26:181-188.

27. Battese GE, Coelli TJ. Model for Technical Inefficiency Effects in a Stochastic Frontier Production Function for Panel Data. Empirical Economics. 1995;20:325-32.

28. Aigner DJ, Lovell CAK, Schmidt P. Formulation and Estimation of Stochastic Frontier Production Function Model. Journal of Econometrics. 1977;1(1) 21-37.

29. Meeusen W, Van den Broeck J. Efficiency Estimation from Cobb-Douglas Production Functions with Composed Error International Economics Review. 1977; 18:435-444.

30. Kaduna State Government. Kaduna State information manual. The Kaduna State Government, Federal Republic of Nigeria; 2012.
Available:http://www.kadunastate.gov.ng

31. NPC. National population commission. population census of the Federal Republic of Nigeria. Census Report. National Population Commission, Abuja; 2006.

32. Asogwa BC, Ihemeje JC, Ezihe JAC. Technical and allocative efficiency analysis of Nigerian Rural farmers: implication for poverty reduction, Agricultural Journal. 2011;6(5):243-251.

33. Battese GE, Malik SJ, Gill MA. An investigation on technical inefficiency of production of wheat farmers in four districts of Pakistan. In: Journal of Agricultural Economics. 1996;47(1):37-49.

34. Obeta ME, Nwabo EC. The Adoption of Agricultural Innovations in Nigeria: A Case Study of an Improved IITA technology Package in Anambra State, in Olukosi J.O, Ogungbile AO and Kalu BA (eds), Appropriate Agricultural Technologies for Resource Poor Farmers. A Publication of the Nigerian National Farming System Research Network. 1999;231-245.

35. Oyekele PS. Resource-use Efficiency in Food Production in Gombe State, Nigeria. An Unpublished PhD; dissertation

submitted to the Department of Agricultural Economics, University of Ibadan; 1999.

36. Zalkuwi JW, Dia YZ, Dia RZ. Analysis of Economic Efficiency of Maize Production in Ganye Local Government Area Adamawa state, Nigeria, Report and Opinion. 2010;2(7).
Available:http//www.sciencepub.net/report, Retrieved 18th June, 2013.

37. Ekong EE. Rural Sociology: An Introduction and Analysis of Rural Nigeria, Uyo: Dove Educational Publication; 2003.

38. Okoye BC, Asumugha GN, Mbanaso Cost and Return analysis of Cocoyam production at National Root crops Research Institute, Umudikwe, Abia state, Nigeria. 2009;17363.
Available:http://mpra.ub.uni-muenchen.de/17363

39. Sharma PV, Staal S, Delgado C, Singh RV. Policy, Technical, Environment and Implication of Scaling-up of Milk Production in India. Annex in Research Report of International Food policy Research Institute (IFPRI), Food and Agriculture Organisation. Livestock Industrialization Project Phase II, Washington D.C., IFPRI; 2003.

40. Oladeebo JO, Oluwaranti AS. Profit efficiency among cassava producers: Empirical evidence from South western Nigeria. Journal of Agricultural Economics and Development. 2012;1(2):46-52.

41. Abdullai A, Huffman W. An Examination of profit inefficiency of rice farmers in Northern Ghana. Iowa 45 International Journal of Agricultural Economics & Rural Development . 1998;1:1-7.

42. Ohajianya DO, CE Onyenweaku. Farm size and relative efficiency in Nigeria: profit function analysis of rice farmers" AMSE Journal of modeling, measurement and control. 2002; 2(2):1-16.

43. Nasiru MO. Microcredit and Agricultural Productivity in Ogun state, Nigeria. World Journal of Agricultural sciences. 2010;6(3): 290-296.

44. Wozniak GD. Joint information acquisition and new technology adoption: Later versus early adoption. Rev. Econ. Stat. 1993;75: 438-445.

45. Ojo MA Mohammed US, Ojo AO, Yisa ES, Tsado JH. Profit Efficiency of small scale cowpea farmers in Niger state, Nigeria. International Journal of Agricultural Economics and Rural Development. 2009;2(2):40-48.

17

Performance Response and Blood Profile of West African Dwarf Goats Fed Shea Butter (*Vitellaria paradoxa*) Leaves Supplemented with Diets Containing Different Levels of Sweet Orange (*Citrus sinensis*) Peels

J. Oloche[1*], O. I. A. Oluremi[2] and J. A. Paul[1]

[1]*Department of Animal Production, University of Agriculture, Makurdi, Nigeria.*
[2]*Department of Animal Nutrition, University of Agriculture, Makurdi, Nigeria.*

Authors' contributions

This work was carried out in collaboration between all authors. Author JO designed the study, wrote the protocol and wrote the first draft of the manuscript. Author OIAO viewed the experimental design and all drafts of the manuscript. All Authors were involved in the management of the experimental animals during the feed trial. Authors JO and JAP managed the analyses of the study. Authors JO and OIAO identified the plants while Authors JO and JAP performed the statistical analysis. All authors read and approved the final manuscript.

Editor(s):
(1) Zhen-Yu DU, School of Life Science, East China Normal University, China.
Reviewers:
(1) Anonymous, Egypt.
(2) Anonymous, Italy.
(3) Anonymous, Nigeria.

ABSTRACT

Nine male grower West African Dwarf (WAD) goats with an average weight of 9.30 kg, aged between 5-7 months were used in a completely randomized design to assess the growth performance and blood profile of WAD goats fed shea butter (*Vitellaria paradoxa*) leaves and supplemented with concentrate diets containing different levels of sweet orange (*Citrus sinensis*) peel meal (SOPM). Three dietary treatments were formulated and compounded to contain 0%, 25%, and 50% SOPM, and were designated T_1, T_2, and T_3 respectively in an eighty-four day feeding trial. Results showed that mean daily weight gain (18.10-27.14 g/day), mean daily feed

Corresponding author: Email: juliangi@yahoo.com;

intake (426.6-462.0 gday/) and final weight (10.20-10.39 kg) were not significantly different (P>0.05) among the treatments. Packed cell volume (PCV), red blood cells (RBC), haemoglobin (Hb), mean corpuscular volume (MCV), mean corpuscular haemoglobin concentration (MCHC) and serum biochemical components showed no significant change among the treatments (P>0.05). This showed that WAD goats fed Shea butter as forage can be supplemented with diets containing up to 50% SOPM without compromising either the growth performance or the health status of the animals.

Keywords: Sweet orange peels; Shea butter leaves; growth performance; blood profile and WAD goats.

1. INTRODUCTION

In Nigeria, ruminant production is faced with problems of inadequate nutrition due to shortage of feed and all year round availability of quality forage. Long period of dry season affects feed availability adversely and this in turn affects development of the goat industry [1]. [2] had earlier reported that the forages are unimproved and low in nutritive values during the wet season while during the dry season proper, they are fibrous, lignified with low in protein values and even in short supply. [3] also reported that, the available forages for most part of the year are low in protein content which leads to marked decrease in voluntary intake and digestibility and substantial weight loss of the animals during this period. In spite of the challenges facing the livestock industry particularly in the third world nations, population pressure has not been on the decline, it has rather been progressively increasing. This increase has in turn subjected the demand for protein supply to intense pressure. [4] reported that the increase in world population especially in developing countries like Nigeria calls for urgent improvement in livestock production, in order to keep pace with the ever increasing human population. Without emphasis, small ruminants have the potentials of alleviating the low protein intake of the people in developing countries like Nigeria [5]. This is achievable if the available low quality feeds are supported by concentrate supplements. It is established that conventional feedstuffs are expensive primarily because of their use in human and monogastric nutrition. However, alternative feedstuffs which are nutritionally viable, cheap and not in high demand by humans are presently been exploited for livestock and indeed for goat production [6]. Sweet orange peels an agricultural by-product obtained from sweet orange fruits is available in Nigeria in large quantities, particularly in Benue state. The peels are usually piled in heaps on refuge dumps and by the roadsides by retailers who peel and sell the fruits for direct

consumption. The peels are easily processed by sun-drying for 48 hours. It is readily crushed into meal when crispy, and has been used in the diet of broiler chickens at low levels without harmful effect [7]. Several authors reported that plane of nutrition, disease, genotype, physiological phases such as lactation [8,9,10] affect blood values, which are indicators of the wellbeing of an animal.

Thus this study was designed to evaluate the performance response and blood profile of WAD goats fed Shea butter (Vitellaria paradoxa) forage supplemented with diets containing different levels of sweet orange (Citrus sinensis) peels.

2. MATERIALS AND METHODS

The experiment was carried out in the Small Ruminant Unit of the Teaching and Research farm, University of Agriculture Makurdi, Benue State, Nigeria. Makurdi is located on latitude 7° 41'N and longitude 38°31' E, 9 metres above sea levels [11]. Peels of sweet oranges were collected from retailers who peel and sell the oranges for direct consumption. The peels were sun-dried for 48 hrs, when it became crispy; it was packed and crushed using a cereal grinding mill. The SOPM was used to replace maize offal at 0 %, 25 %, and 50 % in three dietary treatments designed T_1, T_2 and T_3 respectively. Nine WAD grower goats of about 9.300 kg were purchased from Akwanga local government of Nassarawa State. The Animals were vaccinated against Peste de petite ruminants (PPR) from source, using the PPR vaccine, while Ivermectin was used to check endo and ecto parasites. A week to the arrival of the animals to the farm, the experimental pens were properly washed using disinfectants (Izal), and allowed to dry. Thereafter, the cemented floor was spread with wood shavings which served both as litter materials and beddings for the animals. Upon arrival, the animals were weighed and randomly distributed into three treatment groups of three replicates each, and each animal was a replicate. Each goat was housed in a separate compartment equipped with feed and water troughs. 200 g of concentrate supplement was fed by 8:00 hr daily to the goats and after an hour, the forages were served *ad libitum* by suspending the forages from the roof of the cages to the animals using light ropes. This was to enhance feed in take as well as reduce wastage through trampling. Clean fresh water was served to animals daily *ad libitum*. Animals were weekly weighed to evaluate average weight changes. Average feed in take was calculated by subtracting the quantity of feed remaining from quantity that was fed. On the last day of the experiment, blood samples were collected via the jugular vein into two sets of sample bottles for haematology (with ethylene diamine tetra acetic acid (EDTA) and serum biochemistry (without EDTA). After collection, the blood samples were taken immediately to the laboratory for analysis. The PCV, Hb, RBC and the WBC were determined as reported by [12], the MCV, MCH and MCHC were calculated from the PCV, Hb and the RBC as described by [13]. The serum total protein was determined by the biuret method as described by [14], while the serum albumin was determined by the bromoscresol green method of [15]. The globulin was calculated from the total protein and the albumin, while the SGOT and SGPT were determined calorimetrically. Data collected were subjected to the analysis of variance (ANOVA) as outlined by [16] using the [17] statistical software.

3. RESULTS AND DISCUSSION

The dietary composition of the experimental diets is presented in Table 1. The performance of WAD goats fed the experimental diets is presented in Table 2. Final weights were 10.39 kg, 10.33 kg and 10.20 kg and there were no significant differences among the treatments. Mean daily feed in take ranged between 426.60 g - 462.00 g/day and were similar across the treatments. This perhaps suggests the acceptability of the diets containing the different levels of SOPM to the goats. Observed feed intake values were lower than 525.14-546.26 g/day reported by [18] for WAD goats fed diets containing graded levels of SOPM, but comparable with 402.80 - 446.00 g/day reported by [19] for goats in the humid tropics. Mean daily body weight gain ranged from 18.10-27.70 g and did not show any significant difference ($P>0.05$) among the treatments. This may mean that diets containing SOPM were also adequate for goat production. Observed values were higher than the 6.85 g -20.54 g/l day reported by [20] for WAD goats feed diets containing sweet orange peels. Table 3 shows the effects of sweet orange peel meal on haematology of the experimental goats. There were no significant differences ($P>0.05$) among the treatments in all the haematological indices. The PCV value ranged from 20.67-22.67% and was within the reference values of 21-35% for WAD goats [21]. Red blood cells values were 10.55, 12.20 and 12.10 x 10^6 µ/l for T_1, T_2 and T_3 respectively, this was within normal values for clinically healthy goats,

Table 1. Gross composition of experimental diets

Ingredients	T1 (0%SOPM)	T2 (25%SOPM)	T3 (50%SOPM)
Rice offal	20.00	20.00	20.00
Maize offal	48.80	36.60	24.40
Sweet orange peel meal (SOPM)	0	12.20	24.40
Soya bean (full fat)	28.20	28.20	28.20
Bone ash	2.00	2.00	2.00
Common salt	1.00	1.00	1.00
Total	100.00	100.00	100.00
Calculated			
Crude protein	17.00	16.97	16.94
Crude fibre	15.37	15.41	15.45
Ether extract	11.95	10.82	9.69
Ash	6.76	7.02	7.29
Nitrogen free extracts	48.92	49.78	50.63
Metabolizable energy (kcal/kg)	3333.61	3271.50	3209.04

indicating that the goats were not anaemic. Hb values were between 6.88-7.56 g/dl, this was also normal and similar across the treatments, implying that diets containing SOPM also supported adequate oxygen carrying capacity. The WBC values (16.28-17.33 x 103 µ/l) were also not significantly different (P>0.05) among the treatments, indicating that the animals were not battling with any disease condition. The effect of SOPM on the serum biochemistry of WAD goat is presented in Table 4. None of the biochemical indices was significantly different (P>0.05) among the treatments. The total protein

(59.83-61.37 g/l) and albumin (32.70-35.17 g/l) were normal and similar with the control. This implies that the quality of protein in the diets containing SOPM were not inferior to the control and also that the experimental animals were not in a poor health status. SGOT and SGPT are liver enzymes, the similarity (P>0.05) in the values of SGOT and SGPT with the control treatment shows that observed values for these indices were normal and the function of the liver was not compromised as a result of replacing maize offal with sweet orange peel meal.

Table 2. Performance response of the WAD goats fed experimental diets

Parameters	Experimental diets			
	T_1 (0% SOPM)	T_2 (25% SOPM)	T_3 (50% SOPM)	SEM
Initial weight (kg)	9.43	9.29	9.57	0.42
Final weight (kg)	10.39	10.33	10.20	0.30
Total weight (kg)	0.95	1.04	0.63	0.36
Mean daily weight (g)	27.14	27.70	18.10	10.31
Total forage intake (kg)	11.47	12.35	11.70	0.72
Total concentrate intake (kg)	4.99	3.83	3.22	0.73
Total feed intake (kg)	15.79	16.17	14.93	0.72
Mean feed intake (g)	451.20	462.00	426.00	20.66

Table 3. Effect of sweet orange peel meal on the haematology of experimental goats

Parameters	Experimental diets			
	T_1 (% SOPM)	T_2 (25% SOPM)	T_3 (50% SOPM)	SEM
Packed cell volume (%)	22.67	20.67	20.67	1.85 [ns]
Red blood cells (x 10^6 µ/l)	10.55	12.20	12.10	0.62 [ns]
White blood cells (x 10^3 µ/l)	17.33	16.53	16.28	0.83 [ns]
Haemaglobin (g/dl)	7.56	6.88	6.89	0.62 [ns]
MCV (fl)	21.45	17.31	17.10	2.02
MCH (pg)	7.15	5.77	5.70	0.67 [ns]
MCHC (g/l)	33.33	33.34	33.33	0.01

MCV= Mean corpuscular volume, MCH= Mean corpuscular haemoglobin, MCHC= Mean corpuscular haemoglobin concentration

Table 4. Effect of sweet orange peel meal on serum biochemistry of experimental goats

Parameters	Experimental diets			
	T_1 (0% SOPM)	T_2 (25% SOPM)	T_3 (50% SOPM)	SEM
Total protein (g/l)	61.37	59.83	51.90	6.66 [ns]
Albumin (g/l)	35.17	33.13	32.70	1.24 [ns]
Globulin (g/l)	26.20	24.53	28.10	2.35 [ns]
SGOT (iµ/l)	64.30	174.00	183.00	8.94 [ns]
SGPT (iµ/l)	21.00	19.00	17.67	3.70 [ns]

SGPT=Serum glutamic pyruvic transaminase, SGOT=Serum glutamic oxaloacetic transaminase SEM=Standard error of mean, ns=Not significant.

4. CONCLUSION

From the results of the study, it is concluded that, WAD goats on forages can be fed concentrate supplements containing up to 50% SOPM particularly during the period of feed scarcity to improve feed intake and digestibility as well as minimize production cost without adverse effect on growth and the health status of animals.

COMPETING INTERESTS

Authors have declared that no competing interests exist.

REFERENCES

1. Areegbe AO, Oni AO, Adedeji OY, Falola OO, Saka AA. Performance characteristics and nutrient intake of West African dwarf goats fed cassava leaf hay based diets. Proc. 17th Ann. Conf. Anim. Sc. Ass. of Nigerian (ASAN). 2012;559-562.

2. Babayemi OJ, Bamikole MA, Odunguwa BO. Haematological and biochemical components of West African dwarf goats fed *Tephrosia* bracteolate – based forage. Trop. Anim. Prod. Invest. 2003;6:31-38.

3. Lamidi AA, Aina ABJ, Sowande SO, Jolaosho O. Assessment of *Panicum maximum* (Jacq), Glincia sepium (Jacq) and *Gruelina arborea* (Roxb) based diets as all year round feed for WAD goats. Proc.14th Ann. Conf. Anim. Soc. of Nig. Held on the 13th - 17th Sept. 2010, LAUTH, Ogbomosho, Nigeria; 2010.

4. Okah U. Effect of dietary replacement of maize with maize processing waste on the performance of broiler. In; Ogunji JO, Osakwe W, Ewa VW, Alaku SO, Otuma SO, Nweze BO. Proc. 9th Ann. Conf. Anim. Sci. Assoc. Nig. held at Abakaliki. 2004;2-4.

5. Orayaga KT. Effects of duration of water soaking of sweet orange (*Citrus sinensis*) fruit peel on its chemical; composition and growth performance of broiler starter chicks. Anim. Prod. Res. Adv. 20106(4):311-314.

6. Ukanwoko AI, Ibeawuchi JA, Ukachukwu NN. Growth performance and carcass characteristics of West African dwarf goats fed cassava peel-cassava leaf meal based diet. Proc. 34th Ann. Conf. Nig. Soc. for Anim. Prod. held on the March, 2009, Univ. of Uyo, Uyo. 2009;476-479.

7. Schalm N, Jain NC, Caroll EJ. Veterinary haematology. 3rd edition, Lea and Febiger, Philadelphia, U.S.A; 1975.

8. Zumbo A, Di Rosa AR, Casella S, Piccione G. Changes in some blood haemato-chemical parameters of maltese goats during lactation. J. Anim Vet. Advances. 2007;6(5):706-11.

9. Avondo M, Pagano RI, Guastella AM, Criscione A, Di Gloria M, Valenti B, Piccione G, Pennisi P. Diet selection and milk production and composition in Girgentana goats with different αs1-casein genotype. J. Dairy Res. 2009;76:202-209.

10. Garba Y, Abubakar AS. Haematological response and blood chemistry of Yankasa rams fed graded levels of *Tamarindus indica* leaves. Nig. J. of basic and appl. Sc. 2012;20(1):44-48.

11. Microsoft Encarta Microsoft Corporation; 2008.

12. Olukotun O, Nwaosuibe KE. Macro mineral composition of selected forages edible by ruminants in Afaka Area, kaduna. Proc. 33rd Anim. Conf. of Nig. Soc. Anim. Prod. March, 2008, Coll. of Agric. Sc. O.O.U, Ayetoro, Ogun State, Nigeria. 2008;596-598.

13. Jain NC. Veternary haematology. 4th edition Lea and Babings, Philadelphia. U.S.A. 1986;244.

14. Kohn RS, Allen MS. Enrichment of proteolytic activity relative to nitrogen in preparation from the rumen for *in vitro* studies. Anim. Feed Sci. and Technol. 1995;52(1&2).

15. Doumas BT, Biggs HG. Simple method of albumin determination. Clin. Chem. 1972;7:175.

16. Akindele SO. Basic experimental designs in Agricultural Research. Montem Paper Backs, Akure Nigeria. 1996;25-34.

17. MINITAB Statistical Software. MINITAB Inc. P.A., USA. 1991;10(2).

18. Oloche J, Oluremi OIA, Ayoade JA. Performance of West African Dwarf (WAD) goats fed diets containing graded levels of sweet orange (*Citrus sinensis*) peel meal. Proc. 37th Ann. Conf., Nig. Soc. for Anim. Prod. Held on 18th-21st March 2012. Univ. of Agric. Makurdi, Benue State, Nigeria. 2013;367-370.

19. Abiola MO, Osinowo OA, Abiona JA. Feed intake and weight gain in goats subjected to water restriction during hot season in the humid tropics. Proc. 34th Conf. Nig. Soc. for Anim. Prod. Held at Univ. of Uyo on the 15th-18th March, 2009. Uyo. Nigeria. 2009;487-489.

20. Ngi J. The nutritive potential of sweet orange (*Citrus sinensis*) peel meal for goat feeding. A Ph.D Thesis presented to the department of animal producion, University of Agriculture Makurdi, Benue State, Nigeria. 2014;106-108.

21. 6.Daramola JO, Adeloye AA, Fatoba TA, Soladoye AO. Haematological and biochemical parameters of West African Dwarf goats. Livest. Res. for Rural Develop. 2005;17(8):3-5.

18

Correlation between Intake and Ingestive Behaviour of Confined Holstein-Zebu Crossbred Heifers

R. R. Silva[1*], A. C. Oliveira[2], G. G. P. Carvalho[3], F. F. Silva[4], F. B. L. Mendes[4], V. V. S. de. Almeida[2], L. B. O. Rodrigues[5], A. A. Pinheiro[6], A. P. G. Silva[4] and R. M. do Prado[7]

[1]Graduate Program in Animal Science, State University of Southwest Bahia (UESB), Itapetinga-BA, Brazil.
[2]Department of Animal Science. Federal University of Alagoas (UFAL). Arapiraca Campus. AL Alagoas, Brazil.
[3]Department of Animal Science. Federal University of Bahia (UFBA) Campus Salvador, Bahia, Brazil.
[4]Department of Animal Science, State University of Southwest Bahia (UESB), Itapetinga-BA, Brazil.
[5]Graduate Program in Food Engineering, State University of Southwest Bahia (UESB), Itapetinga-BA, Brazil.
[6]Goiás Agency for Technical Assistance and Rural Extension Agricultural Research (EMATER), Goiânia-GO Brazil.
[7]Graduate Program in Animal Science, State University of Maringá (UEM), Maringá-PR, Brazil.

Authors' contributions

This work was carried out in collaboration between all authors. Author RRS designed the study, wrote the protocol and wrote the first draft of the manuscript. Authors ACO, GGPC, FFS, FBLM, VVSA, AAP, and APGS reviewed the experimental design and all drafts of the manuscript and managed the analyses of the study and performed the statistical analysis. Authors LBOR and RMP performed the translation and correction of English. All authors read and approved the final manuscript.

Editor(s):
(1) Zhen-Yu Du, School of Life Science, East China Normal University, China.
Reviewers:
(1) Anonymous, Universidad Autonoma Del Estado de Mexico. Toluca, Mexico.
(2) Anonymous, University of Nigeria, Nsukka, Nigeria.
(3) Marcelo Tsuguio Okano, Paulista University/State Faculty of Technology of Barueri, São Paulo, Brazil.

ABSTRACT

The aim of the study was to evaluate the correlation between variables referring to feed intake and behaviour of confined Holstein-Zebu crossbred heifers. The experiment was conducted at the dairy

Corresponding author: E-mail: rrsilva.uesb@hotmail.com;

unit of the Southwest Bahia State University, campus Itapetinga, Brazil. Sixteen ¾ Holstein x ¼ Zebu heifers with average of 12 months old and initial weight of 150 kg were randomly distributed into four treatments, with four repetitions. A moderate positive correlation was observed between feeding time and NDF and NFC intake variables. The feeding efficiency of NDF was highly correlated with DMI. The variables referring to time of feeding, rumination and boluses per day were observed to be highly correlated with the intake variables, thus showing great potential to draw up predictive equations.

Keywords: Ethology; intake; nutrition; rest; rumination.

1. INTRODUCTION

The animals are able to express their feeding behavior through metabolic alterations caused by the amount of a certain nutrient intake, however According to [1] the correlation between the intake of nutrients, performance, and animal behavior can serve as extremely relevência for the understanding of metabolic and nutritional aspects of beef production tool, eliminating the need for invasive tests that in many cases not meet the requirements of well-being in force, in which the animals are subjected. Thus, in order to improve the knowledge of daily food intake, it is necessary to study its individual components, which can be described by the number of meals consumed each day, the average duration of meals and consumption speed.

The studying the ingestive behavior is a highly important tool in the evaluation of diets, because through it is possible to acquire knowledge of the possible relationships existing between the plant-supplements-animal interface, allowing us to adjust the feeding management of animals to obtain better productive performance [2].

Studies on the ingestive behaviour of ruminants have increased over the last years [3,4,5], as researchers need to distinguish and recognize the behavioural aspects, which are directly related to the need for understanding the variation from diet responses. These variables are not well understood when individually evaluated, thus, they require behavioural observation. [6], report that the study of ingestive behavior of cattle is an important tool for the development of strategies which support research enabling adjustment of feeding and management techniques to improve the growth performance of the animals.

Despite that not all intake variables are directly related to ingestive behaviour, identifying those that can be highly correlated to it might generate subsidies for the establishment of models that can properly estimate consumption, without the need of invasive techniques. These techniques would alter the feeding behaviour of animals and consequently reflect on their consumption and performance.

The quality of forages, especially the fiber content, is one of the factors directly related to the stimulation of chewing, saliva production, rumen motility and rumen maintenance may interfere with feeding behavior [7], where as the consumption of concentrates and finely ground or pelleted hay is associated with reduced rumination time [8]. The objective of this study was to evaluate the correlation between the variables referring to behaviour and food intake by confined Holstein-Zebu crossbred heifers, which will help the development of models.

2. MATERIALS AND METHODS

2.1 Animal Management and Sampling

The experiment was conducted at the Cattle Farming Unit of Southwest Bahia State University from November 2004 to February 2005. Sixteen ¾ Holstein x ¼ Zebu heifers with average age of 12 months old and initial weight of 150 kg were randomly distributed into four treatments and four repetitions. The animals were kept in individual concrete-floored pens of 2.5 m^2. The experimental period lasted 70 days, with 14 days for adaptation. All observations were made on the last week of the experimental period. The animals were fed a diet consisting of forage to concentrate ratio of a 60:40, based on dry matter content and diets were fed ad libitum. Elephant grass silage was replaced by 5, 10, 15 or 20% cassava bagasse. These levels corresponded to the four treatments, which were formulated to be isonitrogenous. The concentrate contained corn, soybean meal, urea + ammonium sulfate and mineral mix (Table 1).

The chemical analysis of experimental diets and cassava bagasse were made as described by [9], and are presented in Table 2.

Food was provided twice a day at 7:00 a.m. and 04:00 p.m. and water was always available. The amount of food provided was adjusted according to the previous day's consumption, ensuring 5 to 10% leftovers. The amount of food provided and leftovers were daily removed and weighed individually.

Concentrates were sampled on a weekly basis. Silage and leftovers were sampled daily and pooled weekly. All samples (silage, concentrates and leftovers) were pre-dried in a forced ventilated oven at 65°C and ground in a 1 mm mesh sieve for subsequent laboratory analysis of dry matter (DM), neutral detergent fibre (NDF) and non-fibre carbohydrates (NFC).

2.2 Ingestive Behaviour

Feeding, rumination and resting observations were made for 48 consecutive hours, with intervals of five minutes [10]. During night-time, artificial lighting was used. Feeding and rumination efficiencies were estimated as g DM/hour, g NDF/hour or g NFC/hour, ingested or ruminated.

Total chewing time, DM/bolus, NDF/bolus, NFC/bolus, boluses/day, time/bolus, chews/bolus was observed for two days, during the last week of the experimental period [11]. Observations were made during three periods of two hours, during the two observational days. Observations were made from 10:00 a.m. to 12:00 p.m., from 02:00 p.m. to 04:00 p.m. and from 06:00 p.m. to 08:00 p.m., as described by [12]. The collection data was performed with the use of digital timers handled by four observers, on the pre-determined periods, to identify the time spent on each activity.

2.3 Experimental Design and Statistical Analysis

A completely randomized experimental design was used, with four treatments and four replications.

Data analysis consisted of Pearson's linear correlation coefficient between behaviour variables and DM, NDF and ADF intake. The coefficients found had their significance obtained through the Student's t-test, at 5% probability, using the software package Statistical Analysis for Genetic Epidemiology - [13].

3. RESULTS AND DISCUSSION

Data of values of intake and ingestive behaviour on which these correlations were based on the report by [14]. There was no correlation between feeding and dry matter intake (DMI) (P>0.05). As for NDF and NFC variables, there was moderate and positive correlation with the feeding time (P<0.05) (Table 3).

By evaluating the ingestive behaviour of cattle finished in feedlot systems and fed different levels of concentrate, [15]. Noted that the results concerning the time allocated to food consumption was highly correlated with the level of neutral detergent fibre; the correlation was 0.70 (P<0.01).

Table 1. Proportion of ingredients in concentrates (%) on the natural basis

Ingredient	Level of cassava bagasse			
	5	10	15	20
Corn meal	57.0	56.2	55.5	54.7
Soybean meal	36.9	37.7	38.4	39.2
Urea	2.0	2.0	2.0	2.0
Ammonium sulfate	0.3	0.3	0.3	0.3
Calcitic limestone	1.7	1.7	1.7	1.7
Dicalcium phosphate	1.2	1.2	1.1	1.1
Mineral salt[1]	1.0	1.0	1.0	1.0

[1]Composition: 18.5% of calcium, 9% of phosphorus, 0.4% of magnesium, 1% of sulfur, 11.7% of sodium, 30 ppm of selenium, 1500 ppm of copper; 4000 ppm of zinc; 1200 ppm of manganese; 150 ppm of iodine, 150 ppm of cobalt

Table 2. Chemical-bromatological composition (% DM) of experimental diets and cassava bagasse

Item (%)	Level of cassava bagasse				Cassava bagasse
	5	10	15	20	
DM[1]	34.92	38.55	41.82	43.36	87.50
CP[2]	13.55	13.88	13.61	14.07	1.91
NDF[3]	48.97	42.94	38.70	36.79	12.02
ADF[4]	27.49	23.59	20.86	19.91	6.73
EE[5]	3.33	3.09	3.08	2.85	0.60
Ashes	7.67	7.04	6.18	6.33	1.62
OM[6]	92.33	92.96	93.82	93.67	98.38
NFC[7]	26.49	33.05	38.42	39.95	83.85

[1]Dry matter, [2]crude protein, [3]neutral detergent fiber, [4]acid detergent fiber, [5]ethereal extract, [6]organic matter, [7]non-fiber carbohydrate

Information on the correlations regarding behavioural and intake variables is of paramount importance to build equations that are predictive of animal's consumption, which represents the key variable on any nutritional trial. After establishing variables that could be potentially used in predictive models, it may be possible to reduce the use of invasive techniques that compromise the animal's responses.

There was a high correlation between rumination time and NDF intake (P <0.05) and a moderate positive correlation with rumination and DM, and NFC intake (P<0.05). There was a strong negative correlation with resting period and NDF intake (P<0.05), and a moderate positive correlation between the variables referring to resting and intakes of DM and NFC (P<0.05). The NDF intake was closely related to the variables involving rumination activities; hence, the negative correlation of this variable with resting time was already expected. [16], found a negative correlation (P<0.05) between NDF for meals in grams and average daily gain. Because the NDF content of the feeds is directly related to intake, and the latter with performance, as the NDF intake is elevated, the intake by the animals might be limited by the filling effect, which may impair performance. The feed efficiency of DM was significantly (P<0.05) moderately negatively correlated (Table 4) with NDF intake (i.e. it indicates an inverse relationship between these parameters).The feeding efficiency of NDF was positively correlated with all intake variables tested, with the highest correlation with DMI. There was no correlation for the feeding efficiency of non-fibre carbohydrates between the consumption variables.

There was a moderate negative correlation between rumination efficiency of DM and NFC

(P<0.05) with NDF intake. Likewise, these variables were not correlated (P>0.05) with the consumption of DM and NFC (Table 2). In this study, the inclusion of cassava bagasse has improved the efficiency of DM rumination, which explains the low correlation with NDF intake, considering that the inclusion of cassava bagasse led to decreased NDF levels.

Unlike the results observed for the above mentioned variable, the efficiency of NDF rumination had a positive correlation with intake of DM and NFC.

The efficiency of rumination is important to control the use of low digestible food, since when Animals are fed high digestible feed they ruminate larger amounts of food during the normal 8 or 9 hours of rumination, thus allowing a greater intake of food and better growth performance [17].

The total chewing time (TCT) had a moderate positive correlation between intake of DM and NFC (Table 5). The NDF consumption was highly correlated with TCT. This result was expected, considering that the stimulus of chewing activities was promoted by the dietary fibre content. There was a negative correlation (P>0.05) between DM per bolus and intake of DM, NDF and NFC (Table 5). On the contrary, the variable referring to NDF per bolus had a high positive correlation with NDF intake and a moderate positive correlation with NFC intake.

Non-fibre carbohydrate per bolus was not correlated (P>0.05) with intake of DM and NFC, but had a negative correlation (P<0.05) between NDF intake.

The number of boluses per day was highly correlated (P<0.05) between the intake of DM,

NDF and NFC. The high correlation between the evaluated variables presents a strong association between the behavioural aspects related to the ingestion and consumption of feed, which could make possible to predict the intake levels through the use of statistical models. Thus, it would not necessarily require the use of invasive methods, which could lead to misleading results or even influence intake levels. [18], found a positive correlation to the number of boluses and body weight. According to the author this connection between body weight and rumination process variables is associated with the volumetric capacity of the rumen, since it is known that the development of the rumen occurs with increasing age of the animal, and that it is connected to lifting of body weight.

There was a high correlation between time per bolus and NDF intake (P <0.05), whereas the observed correlation between NFC intake was positive moderate.

Chewing per bolus was highly correlated (P<0.05) between intake of NDF and NFC, and a moderate correlation between DMI.

[16], found a positive correlation between rumination and the average daily weight gain of the animals, the greater rumination leads to better utilization of ingested food, which consequently allows increase in food intake improving their performance.

Table 3. Linear correlations between ingestive behavior and intake by ¾ holsteinx ¼ zebu heifer in feedlot

Behavior variables	Intake variables					
	DM intake		NDF intake		NFC intake	
	R	P	R	P	R	P
Feeding	-	-	0.81	0.0001	0.53	0.0159
Rumination	0.56	0.0128	0.97	0.0000	0.68	0.0018
Resting	-0.51	0.0229	-0.94	0.0000	-0.65	0.0032

Table 4. Linear correlations between feeding (Effe) and rumination (Efru) (g/h) efficiencies of dry matter (DM), neutral detergent fiber (NDF) and non-fiber carbohydrates (NFC) intake by ¾ holsteinx ¼ zebu heifer in feedlot

Behavior variables	Intake variables					
	DM intake		NDF intake		NFC intake	
	R	P	R	P	R	P
EffeDM	-	-	-0.47	0.0340	-	-
EffeNDF	0.73	0.0006	0.59	0.0085	0.67	0.0023
EffeNFC	-	-	-	-	-	-
EfruDM	-	-	-0.73	0.0006	-	-
EfruNDF	0.86	0.0000	-	-	0.76	0.0003
EfruNFC	-	-	-0.67	0.0022	-	-

Table 5. Linear correlations between aspects of rumination and intake by ¾ holstein x ¼ zebu heifer in feedlot

Behavior variables	Intake variables					
	DM intake		NDF intake		NFC intake	
	R	P	R	P	R	P
TCT[1]	0.50	0.0230	0.94	0.0000	0.65	0.0033
DM/bolus[2]	-0.47	0.0348	-0.92	0.0000	-0.57	0.0100
NDF/bolus[3]	-	-	0.88	0.0000	0.50	0.0231
NFC/bolus[4]	-	-	-0.73	0.0005	-	-
Boluses/day	0.92	0.0000	0.93	0.0000	0.95	0.0000
Time/bolus	-	-	0.86	0.0000	0.57	0.0130
Chews/bolus[1]	0.68	0.0018	0.94	0.0000	0.81	0.0001

[1]Total chewing time, [2]dry matter/bolus, [3]neutral detergent fiber/bolus, [4]non-fiber carbohydrates/bolus

[19], reported that the efficiency of rumination tends to decrease due to the reduction in the rates of fiber in diets with a higher amount of fiber, yet [14] state that depends on the magnitude of variation of the dietary fiber and the rate of its components.

A moderate correlation to the number of ingested ruminated bolus may be due to a greater time spent cake during rumination, or because the composition of the diet with high percentages of concentrated which have a lower amount of NDF causing the animal decrease number of chews per bolus and, consequently, lower rumination by number of cakes per day [16].

4. CONCLUSION

Based on the results obtained in this study enabled us to verify high correlation between time spent eating, ruminating and daily cakes with consumption variable, showing that there is potential for the development of prediction equations to allow for better utilization and consequently better animal performance .

COMPETING INTERESTS

Authors have declared that no competing interests exist.

REFERENCES

1.   Silva ALN, Silva RR, Carvalho GGP, Silva FF, Lins TOJDA, Zeoula LM, Franco SL, Souza SOM, Pereira MS, Barroso DS. Correlation between ingestive behaviour, intake and performance of grazing cattle supplemented with or without propolis extract (LLOS®). Journal of Agricultural and Crop Research. 2014;2:1-10.

2.   Cavalcanti MCA, Batista AMV, GuimA, Lira MA, Ribeiro VL, Neto ACR. Consumption and Ingestive behaviour of sheep and goats feed Palma Gigante (Opuntia Ficus-Indica Mill) and Palma de- Elephant Ear (*Opuntia* sp.). Acta Scientiarum. Animal Sciences. 2008;30(2):173-179.

3.   Missio RL, Brondani IL, Filho DCA, Silveira MF, Freitas LS, Restle J. Ingestive behavior of feedlot finished bulls fed different concentrate levels in the diet. Brazilian Journal of Animal Science. 2010;39:1571-1578.

4.   Silva RR, Prado IN, Silva FF, Almeida VVS, Santana JúniorHA, Queiroz AC, Carvalho GP, BarrosoDS. Diurnal feeding behaviour of Nellore steers fed increasing levels of supplementation on grazing B. decumbens. Brazilian Journal of Animal Science. 2010;39:2073-2080.

5.   Pereira MJ, Branco JO, Christoffersen ML, Pinheiro, TC. Population biology of *Callinectes danae* and *Callinectes sapidus* (*Crustacea*: *Brachyura*: *Portunidae*) in the south-western Atlantic. Journal of the Marne Biological association of the United Kingdom, Cambridge. 2009;89(7):1341-1351.

6.   Correia BR, Oliveira RL, Jaeger SMPL, BagaldoAR, Carvalho GGP, Oliveira GJC, Lima FHS, Oliveira PA. Ingestive behaviour and physiological parameters of steers feed pies biodiesel. Archivos Zootecnia. 2012;61(233):79-89.

7.   Cardoso E, Carvalho S, Galvani DB, Pires CC, Gasperin BG. Feeding behavior of lambs fed diets containing different levels of neutral detergent fiber. Rural Science, 2006;36:604-609.

8.   Van Soest PJ. Nutritional ecology of the ruminant. 2.ed. Ithaca: Cornell, 1994. 476p

9.   Silva DJ, Queiroz AC. Food analysis (chemical and biological methods).3.ed. Viçosa, MG: Edit. UFV, 2002. 235.

10.   Gary LA, Sherritt GW, Hale EB. Behavior of Charola is cattle on pasture. Journal Animal Science.1970;30:303-306

11.   Fischer, V. Effects of photoperiod, the grazing pressure and diet on the feeding behavior of ruminants. Thesis (Ph.D. in Animal Science).Federal University of Rio Grande do Sul, Porto Alegre. 1996;243.

12.   Bürger PJ, Pereira JC, Queiroz AC, Coelho da Silva JF, Valadares Filho SC, Cecon PR, Casali ADP Ingestive behavior in Holstein calves fed diets containing different levels of concentrate. Brazilian Journal of Animal Science. 2000;29(1):236-242.

13.   Federal University of Viçosa - UFV. SAEG -System statistics and genetic analyzes. Version 8.0. Viçosa, MG; 2000. (CD-ROM).

14.   Silva RR, Carvalho GGP, Magalhães AFFF, Prado IN, Franco IL, Veloso CM, Chaves MA,Panizza, JCJ. Feeding behaviour of crossbred heifers grazing in Dutch. Archivos de Zootecnia. 2005;54:63-74.

15.   Missio RL, Brondani IL, Alves Filho DC, Silveira MF, Freitas LS, Restle J. Feeding behaviour of young bulls finished in a

feedlot, fed different levels of concentrate in the diet. Brazilian Journal of Animal Science. 2010;39(7):1571-1578.

16. Dias DLS, Silva RR, Silva FF, Carvalho G GP, Barroso DS, Carvalho VM. Correlation between performance and ingestive behavior of steers post-weaned on pastures. Acta Scientiarum. Animal Sciences Maringá. 2014;36(1):85-91.

17. Mendonça SS, Campos JMS, Valadares Filho SC, Valadares RFD, Soares CA, Lana RP, Queiroz AC, Assis AJ, Pereira, MLA. Ingestive behaviour of dairy cows fed diets based sugar cane or corn silage. Brazilian Journal of Animal Science. 2004;33(3):723-728.

18. Santana Junior HA, Silva RR, Carvalho GGP, Silva FF, Barroso DS, Pinheiro AA, Abreu Filho G, Cardoso EO, Dias D LS, Trindade Jr G. Correlation between performance and feeding behavior of heifers supplemented to grazing. Semina. Agricultural Sciences. 2013;34:367-376.

19. Farias MS, Silva RR, Zawadzki F, Eiras CE, Lima BS, Prado IN. Glycerin levels for crossbred heifers supplemented in pasture: intake behavior. Acta Scientiarum. Animal Sciences Maringá. 2012;34(1):63-69.

19

Correlation between Intake and Feeding Behavior of Holstein Calves Fed Diets Supplemented with Pellets and Mash

R. R. Silva[1*], A. C. Oliveira[2], G. G. P. Carvalho[1], F. F. Da Silva[1], F. B. L. Mendes[1],
V. V. S. De Almeida[2], L. B. O. Rodrigues[1], A. A. Pinheiro[3], A. P. G. Silva[1],
J. W. D Silva[1] and M. M. Lisboa[1]

[1]Graduate Program in Animal Science, Southwest Bahia State University, Brazil.
[2]Graduate Program in Animal Science, Federal University of Alagoas, Brazil.
[3]Graduate Program in Animal Science, Federal University of Goias, Brazil.

Authors' contributions

This work was carried out in collaboration between all authors. Author RRS Author of the project, collaborated with all stages. Authors ACO and FFDS contributed in the writing. Authors GGPC, FBLM, VVSDA, AAP and JWDS contributed in conducting the experiment and laboratory analysis. Author LBOR contributed in the statistical analysis and translation into English. Authors APGS and MML contributed in writing and correction article. All authors read and approved the final manuscript.

Editor(s):
(1) Ismail Seven, Department of Plantal and Animal Production, Vocation School of Sivrice, University of Firat, Turkey.
(2) Masayuki Fujita, Department of Plant Sciences, Faculty of Agriculture, Kagawa University, Japan.
(3) Zhen-Yu Du, School of Life Science, East China Normal University, China.
(4) Hugo Daniel Solana, Department of Biological Science, National University of Central Buenos Aires, Argentina.
(5) Anonymous.
Reviewers:
(1) Anonymous, Algeria.
(2) Anonymous, Italy.
(3) Anonymous, Hungary.
(4) Anonymous, Nigeria.
(5) Anonymous, Uruguay.

ABSTRACT

The objective was to evaluate the correlation between behaviour al and intake variables of Holstein calves fed pellets and meal. Twelve pure Holstein calves with initial average age of 10 days and weight of 27.5 kg were used; The experimental design used in this study was a 2 x 2 factorial

*Corresponding author: E-mail: rrsilva.uesb@hotmail.com;

design (two types of feed and two feeding stages lactation and post-lactation periods), with six repetitions. Was found significant effects the time allocated to eating, rumination and resting. The data concerning the efficiency of feeding and rumination, total chewing time, number of ruminal boluses, rumination time/bolus and the number of chews per cud/bolus and discretization of the time series. There was a moderate positive correlation between the variable behaviour: dry matter per bolus, number of discrete eating and resting periods and the variable of dry matter intake (DMI). There was a strong positive correlation between the variables of behaviour: total chewing time, NDF/BOLUS, ADF/BOLUS and neutral detergent fiber (NDF) and acid detergent fiber (ADF) intake variables. There was a negative moderate correlation between the behaviour variable TRP (time resting period) and the NDF and ADF intake variables. According to the correlations found in this study, between the behaviour aspects and intake of NDF and ADF can support studies aimed to the formulation of diets supplemented based the understanding on behaviour of food intake by animals.

Keywords: Cattle; nutrition; production; ingestion.

1. INTRODUCTION

The probability that food is ingested depends on a number of factors inherent to food, animal behaviour and environment [1].

Several factors can interfere with the daily activities of the animals: forage characteristics, time of food supply, temperature, space available to animals raised in confinement and others. These factors can bring about changes in the time of ingestion or consumption of food, affect the animal performance and hence the efficiency of the productive system.

Given the above, the intake of forage, milk and supplements by calves should be known through the study of ingestive behaviour in order that the nutritional management can be adjusted; thus, the formulation of more appropriate diets that meet the requirements of animals is needed, considering that it is directly connected with the performance and determines how long the animals will remain at this stage [2].

Over the last years, the need for further information on adequate nutritional management methods and understanding the feeding behaviour of ruminants has generated investments in researches. The generation of these data shall make it possible to develop correlational studies that, through detailed examination, will enable the selection of behaviour al variables to predict the consumption of these animals. By identifying these potential variables, it will be possible to develop predictive models that could accurately estimate the variable of dry matter intake, which in turn is considered essential in nutrition and performance trials. Thus, the aim of this work was to evaluate the correlation between the variables of behaviour and food consumption by Holstein calves fed pellets and meal.

2. MATERIALS AND METHODS

2.1 Animal Management and Sampling

The experiment was conducted at the Cattle Farming Unit of University State Bahia, between November 2004 and February 2005. Twelve Black and White Holstein calves aged 10 days and weighing 27.5 kg, were randomly divided into two treatment groups six animals per treatment. The experimental design used in this study was a 2 x 2 factorial design (two types of feed and two feeding stages lactation and post-lactation). The animals were confined in individual concrete-floored stalls of 2.5 m^2. The length of the trial period was 91 days, whereof 7 days were for the adaptation of animals to a new environment, diet and experimental methods.

A diet consisting of a Sprayfo® milk replacer at 1:8 dilution and concentrate were provided *ad libitum* during suckling.

The concentrates had the same bromatological composition, but one of them was supplied as meal and the other as pellets. The composition of the replacer is shown in Table 1 and that of hay and concentrate is shown in Table 2.

The replacer was daily distributed at 7:00 a.m. in five liter plastic buckets, while the concentrate was provided at the same time, in wood-clad plastic buckets. Water was provided ad libitum in full-time available in automatic shell-shaped container. During the first 60 days of age, each animal daily received four kilograms of milk replacer and concentrate *ad libitum*.

Table 1. Chemical composition of the milk replacer used for feeding calves

Item	Replacer (%)
Dry matter (maximum)	97.0
Crude protein[1] (minimum)	20.0
Lactose[1] (minimum)	45.0
Fat content[1] (minimum)	15.0
Fiber[1] (maximum)	0.4
Ashes[1] (maximum)	8.5
Calcium[1] (minimum)	0.7
Phosphorus[1] (minimum)	0.7

[1] % in dry matter.

Table 2. Proportion of ingredients (%) and chemical composition of concentrate and hay in the diet of calves

Item	% of drymatter
Concentrate[1]	
Milled grains of maize	44.8
Soybean meal	39.23
Wheat meal	9.45
Mineral salt[2]	4.0
Palm oil	2.0
Sodium bicarbonate	1.0
Calcitic limestone	0.24
Chemical composition of feed	
DM	91.18
CP	23.38
EE	3.1
NDF	15.02
ADF	8,09
Calcium	1.0
Phosphorus	0.50
Chemical composition of hay	
DM	92.77
CP	6.30
EE	0.35
NDF	81.50
ADF	51.82

Dry Matter (DM); Crude Protein (CP); Ether Extract (EE); Neutral Detergent Fibre (NDF); Acid Detergent Fibre (ADF), [1] 91.8% DM of natural matter.[2] Ca (233 g/kg), P (80 g/kg), Mg (5 g/kg), Na (48 g/kg), Co (25 mg/kg), Cu (380 mg/kg), I (25 mg/kg), Mn (1080 mg/kg), Se 3.75 mg/kg, Zn 1722 mg/kg, 300.000 U.I of vitamin A/kg, 55.000 U.I of vitamin D/kg, 200 mg of vitamin E/kg

The amount of concentrate provided was adjusted according to the previous day's intake, thus allowing an availability rate of 5-10% as leftovers (for safety). The amount of concentrate

provided was daily recorded and the amount of leftovers was weekly assessed and weighed for every single animal. The animals were weaned at 60 days of age and then started to be fed two Kg concentrate/day and hay made from chopped and pre-dried elephant grass (*Pennisetum purpureum*) (*ad libitum*). Hay supply was monitored in order that 5-10% leftovers were produced. The concentrate feed samples were stored in plastic bags and shortly thereafter ground in a 1mm mesh sieve for subsequent laboratory analysis for Dry Matter (DM), Crude Protein (CP), Ether Extract (EE), Neutral Detergent Fibre (NDF) and Acid Detergent Fibre (ADF) [3].

2.2 Ingestive Behaviour

The animals were subjected to four full-time observation periods (24 hours) by trained observers, for evaluation of their ingestive behaviour, totaling two days of observation per experimental period (lactation and post-lactation periods) [4]. In order to determine the time spent on each activity, data collection was performed using appropriate spreadsheet. At night time, artificial lighting was used. The animals were simultaneously observed at intervals of five minutes, during the 24 hour period, totaling 288 daily observations [5] to identify the amount of time allocated to eating, rumination and resting. The collection and processing of data concerning the efficiency of feeding and rumination (g/h), total chewing time, number of ruminal boluses, rumination time/bolus and the number of chews per cud/bolus using digital timers handled by trained observers were carried out according to the procedures described by [6]. Discretization of the time series was directly made in data collection spreadsheets, including the count of discrete feeding, rumination and resting periods, as described by [7]. The average duration of each discrete period was obtained by dividing the day-time of every activity by the number of discrete periods.

2.3 Statistical Analysis

Data analysis was carried out with the Pearson's linear correlation coefficient between the behaviour variables and the DM, NDF and ADF intakes. The coefficients found had their significance obtained through the Student's t-test at 5% probability, using the software package Statistical Analysis for Genetic Epidemiology – SAEG - Federal University of Viçosa [8].

3. RESULTS AND DISCUSSION

The variables, rumination and resting are not correlated with DMI (P>0.05). According to [9], the time spent by the animals to rest and rumination do not interfere in the consumption of dry matter and neutral detergent fiber, therefore, cannot be used as indicators of consumption actual, indicating that these variables do not correlate. Regarding the intake of NDF and ADF, these variables (P<0.05) showed positive correlation with feeding and strong positive correlation with rumination and resting (Table 3).

In line with the results observed in this study according to results found for [10], have evaluated the ingestive behaviour of confined cattle fed different levels of concentrate and have found that the levels of neutral detergent fiber (NDF) had a correlation of 0.72 with the total resting time.

Regarding the behaviour al variables, number feeding periods (NFP) and number periods of resting (NRP), there was positive correlation with the dry matter intake variable (DMI) (Table 4).

It is therefore possible to associate such occurrence with the dry matter intake (DMI) that can determine a higher concentration of

necessary nutrients to meet growth needs and this consumption can be related to several factors. [11], reporting that the level of concentrate used in the diet, may influence changes in consumption, nutrient digestibility and the performance parameters.

The times per feeding (TFP) and rumination period (TRuP) were highly correlated with NDF and ADF intakes (Table 4). Correlations between the TFP and TRuP with the NDF and ADF intake are probably due to the massive rate of passage of forage, whose limitation is the physical effect caused by the high hay NDF content. When high NDF concentrations are present in the diet, the animal tends to have a greater number of meals and consequently there is direct interference in TFP and TRuP expenses for acquiring and breaking food into smaller particles.

Moderate and negative correlations were observed between TRP (r = -0.58, P = 0.00229) and NDF and ADF intakes. The lower the interest in animals concentrate diet, the greater the consumption of rumination and neutral detergent fiber per unit of time, since in general have a high forage NDF and ADF. The probable explanation for this is that, since the activities are mutually excluding when NDF is increased, the animals

Table 3. Correlation between the sum of variables related to behaviour and food intake by holstein calves

Behaviour variables	Intake variables					
	DM intake		NDF intake		ADF intake	
	r	P	R	P	r	P
Feeding	-	-	0.76	0.0022	0.69	0.0062
Rumination	-	-	0.87	0.0001	0.89	0.0001
Resting	-	-	0.85	0.0002	0.84	0.0003

Dry Matter (DM); Neutral Detergent Fibre (NDF); Acid Detergent Fibre (ADF)

Table 4. Correlation between number and time of feeding, rumination and resting and dry matter (DM), neutral detergent fiber (NDF) and acid detergent fiber (ADF) intake by holstein calves

Behaviour variables	Intake variables					
	DM intake		NDF intake		ADF intake	
	r	P	R	P	r	P
NFP[1]	0.76	0.0022	-	-	-	-
NRuP[2]	-	-	-	-	-	-
NRP[3]	0.53	0.0376	-	-	-	-
TFP[4]	-	-	0.82	0.0005	0.79	0.0010
TRuP[5]	-	-	0.85	0.0003	0.89	0.0001
TRP[6]	-	-	-0.63	0.0135	-0.58	0.0229

[1]number feeding period, [2]number rumination period, [3]number resting period, [4]time feeding period, [5]time rumination period, [6]time resting period

have to feed and ruminate for longer periods, which results in a reduction in time per resting period.

The efficiency of feeding and rumination of DM was moderately positively correlated with DM (Table 5). As the food was provided in the form of pellets or meal, the observed time of ingestion of the pellets was observed to be shorter than that regarding the meal. That has occurred in both feeding phases and may be explained by the fact that the more concentrated the diet, the higher its intake and efficiency.

The presence of concentrated supplement in animal diet contributes to an increase in efficiency of feeding and efficiency of rumination, [12], report that food concentrates require less time to ruminated forages when compared to the thus animals can ruminate larger amount of dry matter per unit time (kgMS.hora-1).

Unlike the results for DM intake, NDF intake was not correlated (P>0.005) with the efficiency of feeding and rumination regarding DM, NDF and ADF. Yet, regarding ADF consumption, there was only moderate and positive correlation with the feed efficiency of DM (Table 5). According to [13] the feed efficiency used by animals to acquire food is related to the time allocated for consumption of food and to the specific weight of the food consumed.

There was a moderate and positive correlation between dry matter per ruminated bolus (DM/BOLUS) and dry matter intake (DMI) as shown in Table 6.

Considering that the dry matter intake does not fully reflect the amount of fractions with lower digestibility potential and that it may vary depending on the levels of non-structural carbohydrates of NDF and ADF levels, dry matter intake truly do not show a high correlation with the variable under study. Understanding nutrient intake through the study of behaviour al variables would provide less invasive and

Table 5. Correlation between the efficiency of feeding (Effe) and rumination (Efru) (g/h) and dry matter (DM), neutral detergent fiber (NDF) and acid detergent fiber (ADF) intake by holstein calves

Behaviour variables	Intake variables					
	DM intake		NDF intake		ADF intake	
	r	P	R	P	R	P
EffeDM	0.65	0.0106	-	-	-	-
EffeNDF	0.85	0.0003	-	-	-	-
EffeADF	0.82	0.0005	-	-	0.52	0.0411
EfruDM	0.58	0.0233	-	-	-	-
EfruNDF	0.97	0.0000	-	-	-	-
EfruADF	0.95	0.0000	-	-	-	-

EffeDM = efficiency feeding dry matter; EffeNDF = efficiency feeding neutral detergent fiber; EffeADF = efficiency feeding acid detergent fiber; EfruDM = efficiency rumination dry matter; EfruNDF = efficiency rumination neutral detergent fiber; EfruADF = efficiency rumination acid detergent fiber

Table 6. Correlation between the sum of variables related to behaviour and food intake by holstein calves

Behaviour variables	Intake variables					
	DM intake		NDF intake		ADF intake	
	r	P	R	P	R	P
TMT[1]	-	-	0.80	0.0009	0.80	0.0010
DM/BOLUS[2]	0.67	0.0085	-	-	-	-
NDF/BOLUS[3]	-	-	0.99	0.0000	0.98	0.0000
ADF/BOLUS[4]	-	-	0.98	0.0000	0.97	0.000
BOLUSES/DAY	-	-	0.98	0.0000	0.94	0.0000
TIME/BOLUS	-	-	0.63	0.0137	0.61	0.0184
CHEWING/BOLUS	-	-	0.62	0.0150	0.60	0.0192

[1]total chewing time, [2]dry matter (DM) /bolus, [3]neutral detergent fiber (NDF) /bolus, [4]acid detergent fiber (ADF) /bolus

stressful solutions. Such information is extremely important for the traditional nutritional studies, considering that they interfere with the environment where the animals are raised, causing a series of inconveniences that can disturb their biological functioning. This study has identified a high positive correlation between the variables of behaviour al aspects such as total chewing time (TMT), neutral detergent fiber per ruminated bolus (NDF/BOLUS), acid detergent fiber per bolus (ADF/BOLUS), daily ruminated boluses, time per ruminated bolus, chews per ruminated bolus and NDF and ADF intakes.

The high correlation between these variables shows the strong association between the behaviour al aspects related to food intake. For a long time, many researchers have put into question the actual usefulness of behaviour al studies related to food intake. The information contained in this publication can potentially guide innovative researches aimed at the use of behaviour al variables as a tool for guiding the development of new predictive food intake methods for both confined and in field cattle.

4. CONCLUSION

According to the correlations found in this study, between the behaviour aspects and intake of NDF and ADF can support studies aimed to the formulation of diets supplemented based the understanding on behaviour of food intake and may cause positive results for animal performance.

COMPETING INTERESTS

Authors have declared that no competing interests exist.

REFERENCES

1.   Pereira ES, Mizubut IY, Ribeiro ELAA, Villarroeli ABS, Pimentel PG. Intake, apparent digestibility of nutrients and feeding behaviour of Holstein cattle fed hay diets containing Tifton 85 with various particle sizes. Brazilian Journal of Animal Science. 2009;38:190-195.

2.   Silva RR, Silva FF, Mendes FBL, Almeida VVS, Carvalho GGP, Lisboa MM. Study of correlations between discrete periods of Behaviour with the variables of food intake

by Holstein calves supplemented with pelleted and mash. Proceedings ... 48th Annual Meeting of the Brazilian Society of Animal Science: The Development of Animal Production and Responsibility Challenges Facing New Belém – PA; 2011.

3.   Silva DJ, Queiroz AC. Food analysis (chemical and biological). Viçosa: Federal University of Viçosa. 2002;235.

4.   Fischer V. Effects of photoperiod, the grazing pressure and diet on the feeding behaviour of ruminants. Thesis (Ph.D. in Animal Science) - Federal University of Rio Grande do Sul, Porto Alegre. 1996;243.

5.   Gary LA, Sherritt GW, Hale EB. Behaviour of Charolais Cattle on Pasture. Jounal Animal Science. 1970;30:303-306.

6.   Bürger PJ, Pereira JC, Queiroz AC, Coelho da Silva JF, Valadares Filho SC, Cecon PR, Casali ADP. Ingestive behaviour in Holstein calves fed diets containing different levels of concentrate. Brazilian Journal of Animal Science. 2000; 29(1):236-242.

7.   Silva RR, Silva FF, Veloso CM, Aguiar MSM, Carvalho GGP, Almeida VS, Santos CB, Dutra GS, Matos RS. Assessment of Ingestive Behaviour of ¾ Holstein x Zebu heifers fed elephant grass silage plus 10% cassava meal. Methodological Aspects. In: 40th Annual Meeting of the Brazilian Society of Zootecnia, 40, 2003, Santa Maria. Proceedings ... Santa Maria: SBZ; 2003.

8.   Federal University of Viçosa. SAEG - System statistics and genetic analysis. Version 8.0. Viçosa. User manual. 2000;142.

9.   Pinto AP, Marques JA, Abrahão JJS, Nascimento WG, Costa MAT, Lugão SMB. Behaviour and Ingestive Efficiency of Young Bulls in Feedlot with Three Different Diets. Archivos de Zootecnia. 2010;59 (227):431.

10.   Missio RL, Brondani IL, Filho DCA, Silveira MF, Freitas LS, Restle J. Ingestive Behaviour of feedlot finished bulls fed different concentrate levels in the diet. Brazilian Journal of Animal Science. 2010; 39:1571-1578.

11.   Mateus RG, Silva FFD, Ítavo LCV, et al. Supplements for rearing of Nellore cattle in the dry season: performance, consumption

and nutrient digestibility. Acta Scientiarum. Animal Sciences. 2011;33 (1):87-94.

12. Mendes Neto J, Campos JMS, Valadares Filho SC, Lana RP, Queiroz AC, Euclydes RF. Feeding behavior of dairy heifers fed citrus pulp in substitution of Tifton 85 hay

Brazilian Journal of Animal Science. 2007; 36(3):618-625.

13. Van Soest PJ. Nutritional ecology of the ruminant. 2.ed. Ithaca: Cornell University Press. 1994;476.

20

Milk Production and Supply Chain in Peri Urban Areas of Jhang Pakistan

Nadeem Abbas Shah[1*], Nowshad Khan[1], Raees Abbas[2], Muhammad Hammad Raza[2], Babar Shahbaz[2], Badar Naseem Siddiqui[3], Farhat Ullah Khan[1] and Shafique Qadir Memon[1]

[1]*Department of Agricultural Sciences, Allama Iqbal Open University, Islamabad, Pakistan.*
[2]*Institute of Agriculture, Extension and Rural Development, University of Agriculture, Faisalabad 4000, Pakistan.*
[3]*Pir Meher Ali Shah Arid Agriculture University, Rawalpindi, Pakistan.*

Authors' contributions

This work was carried out in collaboration between all authors. Authors NAS and NK designed the study, wrote the protocol and wrote the first draft of the manuscript. Authors RA and BS reviewed the experimental design and all drafts of the manuscript. Authors MHR and BNS managed the analyses of the study. Authors FUK and SQM performed the statistical analysis. All authors read and approved the final manuscript.

Editor(s):
(1) Anonymous.
Reviewers:
(1) Anonymous, Haramaya University, Ethiopia.
(2) Anonymous, King Abdulaziz University, Saudi Arabia.
(3) Andell Edwards, Animal Science, University of Trinidad and Tobago, Trinidad and Tobago.
(4) Theodros Tekle, Animal Production Course Team, College of Veterinary Medicine, Mekelle University, Ethiopia.
(5) Anonymous, University of the Free State, Namibia.
(6) Anonymous, Paulista University, Brazil.

ABSTRACT

The peri urban areas are facing the challenge of poverty, hunger, unemployment, pollution and ineffective utilization of natural resources. Present paper focuses on milk production and its supply chain in the peri urban areas of Jhang district, Punjab province, Pakistan. Quantitative and qualitative data obtained through structured interview schedule, focus group discussion and personal observation are analyzed through statistical package for social sciences (SPSS) and content analysis technique. Results of the milk supply chain revealed that an overwhelming

Corresponding author: E-mail: nabbasuaf@yahoo.com;

majority of the respondents have no proper place for animals keeping. It was observed that in the peri urban areas most of the milk producers sell milk directly to their neighbors and also they sell to the *gawalas* Fellow farmers are the main source of information for about 32% of the respondents. Cows and buffaloes are major milk producing animals in the study area and milk is mostly sold through Gawalas as a middlemen. Therefore, interventions are required which focus on the control of diseases and marketing aspects of milk production in the peri urban areas of Jhang district, Pakistan.

Keywords: Milk marketing; milk production; peri urban; Pakistan.

1. INTRODUCTION

The word 'peri urban' refers to "the margins of big cities" [1]. Peri urban areas are situated within the municipal areas of a country but are often outside the formal urban boundaries where both agricultural and non-agricultural activities exist simultaneously, though the agricultural and rural characteristics are gradually replaced by urban lifestyles. Peri urban areas can also be defined as a region just around the built up territory of a city, its edge, the 'rural-urban fringe' where city and country land uses go beyond [2]. Peri urban boundary has a place that has a large impact on people. Due to the flow of goods, resources, people and capital to peri urban areas, the sustainability of cities and rural areas is affected. The peri urban areas are facing the challenge of poverty, hunger, unemployment, pollution and ineffective utilization of natural resources [3]. The combination of urban and rural activities with negligent regulations, industrial and urban waste, and intensive agriculture releases of toxic chemicals combined with poor waste and sanitation management often makes peri urban areas much polluted place [4]. Douglas [5] stated that the peri urban areas are facing significant challenges of access to land, problems related to soil and water quality gradually more polluted by peri urban waste, urban waste disposal and industrial activities.

Livestock sector is thought to be one of the most important sectors in the economy of peri urban areas. Earlier it was a traditional practice but now livestock are mostly reared for commercial purposes. It plays very important role in enhancing the health and food security status in developing country. Livestock holding is considered as an important strategy for poverty reduction in rural communities of Pakistan. The livestock accounts for approximately 55% of the agriculture value added and 11% to Gross Domestic Product (GDP) during 2012-13 (Govt. of Pakistan, 2013) [6]. Livestock commodities are very important for farmers because it provides

independent of land and free ownership of assets. Livestock products and byproducts provide fast cash at the time of need and food. The dairy products like fresh milk, butter, *lassi* and *desi ghee* have important role in improving food security condition of the people [7].

Livestock products, especially dairy products provide protein, vitamin A, carbohydrates, and calcium which can make a significant contribution to human nutrition in developing countries [8]. Milk is very important commodity of rural and peri urban areas providing food requirements to the dwellers. Milk produced in peri urban areas is either used for food in the household or sold to earn some money. The earned money is used to buy other commodities including food items for the family. It significantly enables the milk producers to be prosperous [9]. Sharif et al. [10] noted that 80% milk is produced by rural areas, while 20 to 25% produced by peri urban areas and cities are the low producers of milk. The same author also further stated that about 90% of the sold milk comes from small farmers whereas the remaining is supplied by commercial farms. The traditional milk marketing channel involves the collection of milk from farmers in remote areas by the village *dodhi*. The unprocessed milk reaches the final consumer through a variety of peoples like through middleman, collection centers, milk shops, sweet shops, and *gawalas*. Transportation costs for moving milk from small towns or transaction points on main highways, to cities, ranges between Rs: 10-50 (Pakistani Rupees) depending on the size of the utensil, mode of transportation and distance involved.

Milk is very important part of the daily food of Pakistan but pure milk is not easily available. Mostly milk is supplied to big cities from peri urban areas through different ways and means. Milk producers in peri urban areas are facing many constraints regarding rearing livestock and milk marketing. The main concern of the research work is to identify the problems of producers and supply chain. Therefore, this study was conducted with the following objectives: to identify the problems regarding livestock rearing, assess the animal's diseases awareness level and to observe the milk production and supply chain.

2. MATERIALS AND METHODS

The study was conducted in peri urban areas of Jhang city. It is situated in Central Punjab, Pakistan. It is the oldest city of the Central Punjab. The sample of population was limited to 120 peri urban dwellers selected through simple random sampling technique. Three peri urban areas namely Hasnana, Kotsai Singh and Pukay Wala are under peri urban distribution. From each peri urban areas 40 peri urban dwellers were selected as respondents making sample size of 120. In order to collect the required information, a validated and reliable interview schedule was used for quantitative data. In order to collect qualitative data, focus group discussion and personal observation were used. The quantitative data were statistically analyzed by using computer software Statistical Package for Social Sciences (SPSS) [11] and qualitative data through content analysis technique.

3. RESULTS AND DISCUSSION

3.1 Age of the Respondents

Table 1 shows that about 32.5% of the respondents were up to 40 years of age, 20.8% of the respondents were up to 41-50 years of age. However, 32.5% of the respondents were above the age 50-60 years and 14.2% above 60.

Table 1. Distribution of the respondents according to their age

Age group	Frequency	Percentage
<= 40	39	32.5
41 – 50	25	20.8
51 – 60	39	32.5
61+	17	14.2
Total	120	100.0

It is generally believed that with the increase in age, the individual become mentally mature and takes rational decisions and therefore, age can be one of the important factors affecting the behavior of the respondents [12].

3.2 Education Level of the Respondents

The 23% respondents in the peri urban areas were illiterate because they have no facilities for education. Qualitative data as shown in Table 2, revealed that most of the people in these areas belong to poor families so they have no access to education.

Table 2. Distribution of the respondents according to their education

Education level	Frequency	Percentage
Illiterate	28	23.3
Primary	26	21.7
Middle	41	34.2
Above	25	20.8
Total	120	100.0

Education can be defined as the process of developing knowledge, wisdom and other desirable qualities of mind, character and general competency, especially by a source of formal instructions.

3.3 Animals Rearing

Livestock enterprises are particularly important for the landless and small farmers because livestock provides an alternative form of asset ownership, independent of land. Finally, for subsistence farmers, livestock products like fresh milk and butter or *desi* ghee provide food security for the family and help to meet nutritional requirements in terms of calories and protein. Some livestock like donkey, bulls and camels are also used for the transportation purposes in these areas. Distributions of the respondents according to the number of animals are given in Table 3.

With regard to the production and use of livestock products; cows and buffaloes were mostly reared for milk production. In the study area, a single cow produces about 6-8 liter milk/day while 10-12 liter/day milk was produced

Table 3. Distribution of the respondents according to the number of animals

No. of animals	Yes (%)	No (%)
Cows	20	80
Buffaloes	25	75
Sheep	12	88
Goat	16	84
Hen	43	57

by buffaloes. About one third of the total milk produced was used for domestic purpose and remaining amount was sold out to nearby milk shop, *gawalas*, hotels etc. They also produce milk by-products like "dasi ghee and butter" and sell out for earning. Some peoples also have a small flock of sheep and goat. Mostly these are reared for meat purpose but some people use goat milk for domestic use. peri urban dwellers mostly sell out goat and sheep on *"Eid-ul Azha"* because at this occasion they get much better profit than other days. However, some people also sell out in *"Maveeshi Mandi"* (livestock market) when they are in drastic condition.

3.4 Proper Barn for Animal Keeping

The data from Table 4 reveals that more than 50% of respondents have no proper barn for animal while they are also rearing animals get money and products from their animals.

Table 4. Distribution of respondents according to the proper place of animal

Proper barn for keeping animal	Frequency	Percentage
Yes	48	40
No	72	60
Total	120	100

3.5 Animal Diseases Awareness

Animal's diseases are the major problem in peri urban and also in rural areas. Due to these diseases, they are facing challenges for production. As indicated in Table 5 more than 50.8% respondent has awareness regarding animal diseases whereas the remaining 49.2% has no information regarding diseases of animal.

Table 5. Distribution of respondents regarding animal diseases awareness

Animal diseases awareness	Frequency	Percentage
Yes	61	50.8
No	59	49.2
Total	120	100

3.6 Information Sources

The peri urban dwellers get their latest information regarding livestock and agriculture from different sources. Most of the respondents

(32.5%) get their information from their neighbor farmers Table 6. About 20.0% of the respondents have extension department as main source of information regarding the livestock and agriculture, while 11.7, 6.7 and 17.5 respondents have T.V, radio and other sources of information, respectively. 15% are not able to get any information from any source. Newspaper has no role to disseminate agricultural information in the peri urban areas.

Table 6. Distribution of respondents according to source of information

Source of information	Frequency (out of 120 respondents)	Percentage
Neighbor farmers	39	**32.5**
Extension department	24	**20**
Other sources (Internet)	21	**17.5**
Not get information	18	**15**
TV	14	**11.7**
Radio	0	**6.7**
Newspaper	**0**	**0**

3.7 Milk Supply Chain in Peri Urban Areas of Jhang District

Milk is produced from dairy farmers in variable quantity depending on number of milch (milking) animals and better management practices. Milk produced in the peri urban areas is either directly bought by the neighbors or the milk producer sell their produced milk to the *gawalas* (called as middle man) after full filling their own requirements. The neighbors directly getting milk from the producers are satisfied with the purity of milk although it cost is higher than market price. Further *gawalas* sell fresh milk to the end users and also to the milk shops, hotels and sweet shops, these milk shops process milk by boiling, chilling, making cream, yogurt butter and *khoae*. When the milk is transported through *gawalas* milk is not pure because *gawalas* might add water to increase the milk quantity and also use other chemicals to have thick milk which are very harmful for health. Mostly in the hotels milk is used for tea and *lassi* (yogurt shake). In the sweet shops milk is either used for making sweets or packed for daily usage. From these shops milk and milk by products are directly supplied to the end users or consumers. Detail of milk supply chain is given in Fig. 1.

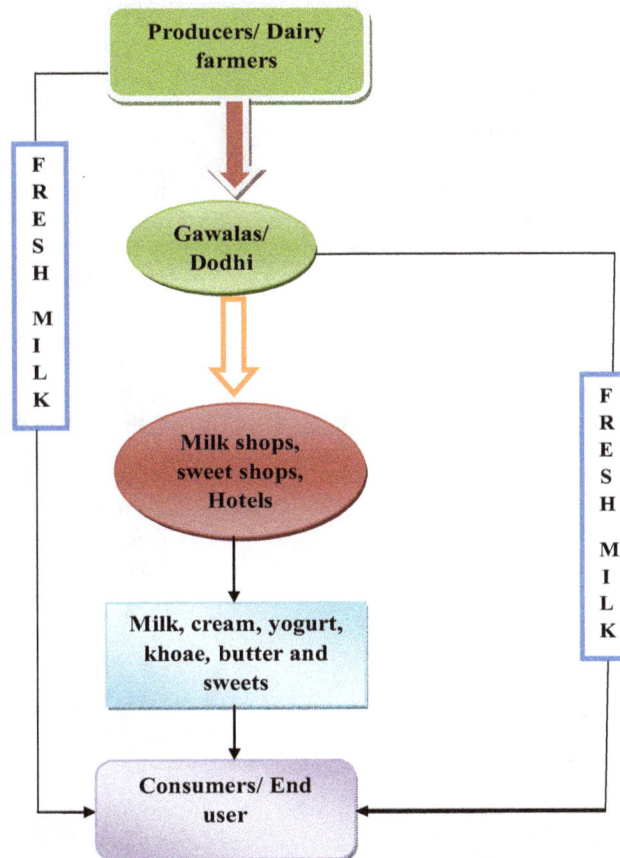

Fig. 1. Milk supply chain in the peri urban areas of Jhang district

4. CONCLUSION AND RECOMMENDATIONS

It is concluded that livestock have massive share in every country like Pakistan's economy and it is also a good source of food and cash income. However, farmers do not have animal sheds and have little knowledge about credible information source(s) regarding diseases of animals and marketing of the milk. The role of middlemen (gawala) was dominant in milk marketing chain. Therefore government should focus on diseases and marketing aspects of milk production in peri urban areas of Jhang district, Punjab, Pakistan.

ACKNOWLEDGEMENTS

We are thankful to our teachers and friends for valuable discussion. One of the authors, Raees Abbas acknowledges the Higher Education Commission (HEC) of Pakistan for the award of Indigenous 5000 PhD scholarship program and University of agriculture Faisalabad (UAF), Pir Meher Ali Shah Arid Agriculture (PMAS) University Rawalpindi and Allama Iqbal Open University (AIOU) Islamabad Pakistan for providing research facilities.

COMPETING INTERESTS

Authors have declared that no competing interests exist.

REFERENCES

1. Narain V, Nischal S. The peri urban interface in Shahpur Khurd and Karnera, India. Environment and Urbanization. 2007;19(1):261-273.
2. Adell G. Theories and models of the peri urban interface: A changing conceptual landscape; 1999.
3. Mycoo M. Sustainable livelihoods in the peri urban interface: Approaches to sustainable natural and human resource use, USA. 2006;1(1):134-148.

4.   Singh S, Kumar M. Heavy metal load of soil, water and vegetables in peri urban Delhi, Envir. Monitoring and Assessment. 2006;120(1):79-91.

5.   Douglas I. Peri urban ecosystems and societies: transitional zones and contrasting values. In McGregor D, Simon D, Thompson D (eds) The peri urban Interface: Approaches to Sustainable Natural and Human Resource Use, Earthscan, London. 2006;12(1):18–29.

6.   Govt. of Pakistan. Economic Survey: Economic Advisor's wing, Finance Division, Islamabad; 2013.

7.   Burki AA, Khan MA, Bari F. The State of Pakistan's Dairy Sector: An Assessment. Working Paper No. 05-34. Center for Mgt and Eco. Res., Lahore Univ. of Mgt. Sci. Lahore; 2005.

8.   Ahmed S. K. E. Mohamed, Forthcoming BG. Evidence from the highlands of Ethiopia. Socio-economic and Policy Research, Livestock Policy Analysis Program: Inte. Livestock Research Inst. (ILRI), Ethiopia. 2003;1(1):1-47.

9.   MoA (Ministry of Agriculture). The role of village dairying co-operative in dairy development, Prospects for improving dairy in Ethiopia. 1998;1(1):1-99.

10. Sharif M, Malik W, Hashmi NI, Farooq U. Action plan for livestock marketing systems in Pakistan, social sciences division. Pakistan Agriculture Research Council, Islamabad. 2003;13(1):87-128.

11. Statistical Package for Social Sciences (SPSS).

12. Amir J. An investigation in to the adoption of broiler production / management practices by poultry farmers in tehsil Samundari. Unpublished Master's Thesis, Dept. of Agri. Ext., Univ. of Agri. Faisalabad; 2003.

21

Cointegration and Market Integration: An Application to Hides and *'Pomo'* Markets in Nigeria

O. Yusuf[1*] and J. O. Olukosi[1]

[1]*Department of Agricultural Economics and Rural Sociology, Institute for Agricultural Research, Ahmadu Bello University, Zaria, Nigeria.*

Authors' contributions

This work was carried out in collaboration between both authors. Author OY designed the study, performed the statistical analysis, wrote the protocol and wrote the first draft of the manuscript. Author JOO read and approved the final manuscript. Both authors read and approved the final manuscript.

<u>Editor(s):</u>
(1) Frantisek Brazdik , Charles University in Prague, Czech Republic.
(2) Hassen B. Hussien, Department of Agricultural Economics, Wollo University, Ethiopia.
(3) Vincenzo Tufarelli, Department of DETO - Section of Veterinary Science and Animal Production, University of Bari "Aldo Moro", Italy.
<u>Reviewers:</u>
(1) Anonymous, Ethiopia.
(2) Huijian Dong, College of Business, Pacific University, Oregon, USA.
(3) Anonymous, Russia.
(4) K.K. Bolarinwa, Department of Agricultural Administration Federal University of Agriculture Abeokuta, Nigeria.

ABSTRACT

Aim: To investigate long run and short run relationship between the prices of hides and *'pomo'* in Kano and Lagos markets in Nigeria, using cointegration techniques.
Study design: The study made use of purely secondary data the prices of hides for the past twenty one years (1988 – 2009) in Kano State and the prices of *'pomo'* for the same number of years in Lagos State to see whether the two markets are cointegrated or fragmented.
Place and Duration of Study: From the Federal Ministry of Agriculture and Natural Resources, Livestock and Veterinary Unit, Gwale, Kano and Federal Ministry of Commerce and Industry, Lagos, in July, 2011.
Methodology: Data on prices of hides and *'pomo'* over a period of twenty one years (1988 – 2009) were used. Unit root test and Johansen cointegration test were carried to find out whether the markets in the two locations are cointegrated or fragmented.
Results: The result showed that there was no cointegration between the two markets. The result of similar trend in the prices of hides and *'pomo'* in the two markets may be attributed to other

---

*Corresponding author: Email: oziyusuf@gmail.com, oyusuf@abu.edu.ng;

factors such as inflation and festivities.

Conclusion: There is no long run relationship in the price of hides and *'pomo'* in the two markets. Therefore, there should be government's commitment to policies that can reduce inflation rate so that prevailing market prices of hides and skins can be stable in different market locations.

Keywords: Cointegration; market integration; hides; pomo; Nigeria.

1. INTRODUCTION

Hides and skins are the main export income generators from the livestock sector in Africa. In the year 2000, Africa's share of total world production was only 5% of bovine hides, 14% of goat and kidskins and 8% of sheep and lambskins [1]. [2] discovered that hides and skins are obtainable in commercial quantities from abattoirs located all over the country, especially in the Northern states of Nigeria. Finished leathers on the other hand are obtainable from tanneries and also from traditional tanners, both of which are also located in the Northern part of the country. According to [3] a large proportion of hides and skins from slaughters that take place in registered abattoirs and slaughterhouses, are eaten by human beings as meat supplements and delicacies usually called *'pomo'* in Nigerian parlance. While the hides are flayed before reaching the *'pomo'* eaters, the skins particularly, those of goats are usually roasted together with the entire animal. This serves as special delicacies in many Nigerian homes and restaurants.

'Pomo' is a hide that has undergone some processing such as roasting and boiling for human consumption as meat supplement and delicacy. Usually, it is cowhides that are used as *'pomo'* in Nigeria. The process of removing hairs from hides to become *'Pomo'* is traditionally done by roasting or by tenderizing the hides in hot water, followed by shaving with razor blade. It is then boiled in water to soften the hides before it is used in soup or stew. It has low nutritional value but many Nigerians enjoy eating it as delicacy especially in the western part of the country.

Cointegration has been used to measure movement in prices from one location to another to infer the conduct of a firm [4]. It is an econometric method that has been applied successfully to integration questions in marketing and financial studies [5]. It is an econometric property of time series variables. If two or more series are themselves non- stationary, but a linear combination of them is stationary, then the series are said to be cointegrated [6]. We need the two price series to be cointegrated to represent long run equilibrium relationships implied by economic theory. Also, many univariate economic time series appear to be integrated of order one [1 (1)] series which is a requirement for cointegration [7].

The integration test concerning the stationarity of any time series can be made using several techniques. Stationarity means that the stochastic properties of a time series (i.e the mean, the variance of the mean and the covariance of the mean with values of time series), are stationary and do not vary with time [8]. For price series with trends, one should test the individual series for integration and between the series for cointegration before modelling relationships between the series. Most economic time series are not stationary because, for example, the mean of the series changes with time, if only because of inflation or seasonality [5].

Several works have been conducted on crops and livestock using cointegration techniques. Information on market integration of hides and *'pomo'* at different locations in Nigeria has been very scanty. This study therefore attempts to analyse the spatial price variations in hides and *'pomo'* markets in Nigeria. It is also hoped that information from this study will help the policy makers and researchers of hides and skins in understanding the price movement of hides and *'pomo'* in Nigeria. It will add to the existing knowledge on market integration of hides and *'pomo'* in Nigeria.

1.1 Objectives of the Study

The broad objective of this study is to analyse cointegration and market integration in hides and *'pomo'* Markets in Nigeria. The specific objectives are to:

1. Determine the trends in the Prices of Hides in Kano and Lagos Markets (1988 – 2009)
2. Ascertain stationarity in the two markets
3. Determine whether the prices of hides and *'pomo'* in the two market are co-integrated

1.2 Conceptual Framework and Literature Review

Researches conducted in the late 70's and early 80's on time series data analysis [9,10,11,12,] had led to serious developments in the specification of time series data particularly the non stationary series. Macroeconomic time-series such as income and prices data are mainly non stationary data which need to be made stationary before cointegration test can be carried out. This is against statistical estimation theory that is based on asymptotic convergence theorems which assume that data series are stationary. However, cointegration procedure measures the degree of price integration in two markets A and B. Two markets are said to be spatially integrated, if prices in a given market adjust to the price shock in the other market [13]. Market integration is tested using the cointegration method, which demanded that the prices of the two markets, say P_{at} and P_{bt} must have the same order of integration and a linear combination exists between these two series, where P_{at} is price in market A time t and P_{bt} is price in market B time t.

In recent time, [14] did a wonderful work in the application of cointegration to determine market integration in rice prices in selected areas in Nigeria. They observed that retail prices appear to be rising faster in Lagos than in any other centre. Indications are that prices of local rice rose more in Lagos than in Abuja and Enugu. [15] applied cointegration to examine the relationship between cocoa production in Nigeria and value of loan granted by Agricultural Credit Guarantee Scheme. It was found that there was no long run relationship between the output of cocoa and value of loans granted. However, [16] employed co-integration techniques in analyzing market integration in tomato and onion among various markets in Jigawa State. He discovered that those markets that were close together were more integrated than those far apart. [4] applied it to determine the price of sorghum and maize in two wholesale markets (rural and urban) in Nigeria. They observed that the markets are integrated and that price signals and information are transmitted smoothly across the markets. [17] applied cointegration to determining market integration in maize markets in Pakistan. It was found that the regional markets of maize have strong price linkages and thus are spatially integrated. [18] In his study tested whether the Law of One Price (LOP) holds for maize and rice spatial markets in Malawi using cointegration

techniques in testing the spatial market integration hypotheses. The results suggest that markets for rice crop with complete price liberalization are more integrated than markets for maize in which the governments still imposes a price band for Agricultural Marketing and Development Corporation (ADMARC).

However, an important aspect of the cointegration analysis concerns the specification of processes of dynamic adjustment. Engle-Granger representation theorem [19] states that if two series are cointegrated, then they will be most efficiently represented by an error-correction specification, and furthermore, if the series are cointegrated, this dynamic specification will encompass any other dynamic specification, including the partial adjustment model.

3. METHODOLOGY

3.1 Data Collection

The study was conducted in Nigeria and the two markets selected were Kano market (where the largest market for the sales of hides in Nigeria is found) According to [20] over 85% of the tanneries in Nigeria are found in Kano State. Lagos market (where was expected to have the largest market for the sales of 'pomo'). Secondary data on the prices of hides for the past twenty one years (1988 – 2009) were gathered from the Federal Ministry of Agriculture and Natural Resources, Livestock and Veterinary Unit, Gwale, Kano. The prices of 'pomo' for the same years were also gathered from Federal Ministry of Commerce and Industry, Lagos, to find out if the price of 'pomo' in Lagos is influenced by the prices of hides in Kano and assess if the two markets are fragmented or co-integrated.

3.2 Analytical Techniques

Cointegration tests were conducted after examining the univariate time series properties of the data to confirm that the price series in the two markets are stationary and integrated of the same order. The first step is the unit root test, that is, test for stationarity of the series. In testing the null hypothesis of non stationarity against the alternative, Augmented Dickey Fuller (ADF) tests were employed. The ADF tests the null hypothesis that a series (P_t) is non-stationary and must be differenced at least once before stationarity is achieved. The equation of ADF is

$$\Delta P_t = \beta_0 + \beta_1 P_{t-1} + \gamma_t + \sum_{k=1}^{n} \delta_k \Delta P_{t-k} + e_t \quad (1)$$

Where $\Delta P_t = P_t - P_{t-1}$, $\Delta P_{t-k} = P_{t-k} - P_{t-k-1}$, k = 1, 2, 3, ...n, and P_t, P_{t-1} , P_{t-k}, P_{t-k-1} are the respective prices at times t, t-1,t-k and t-k-1. While β_0, β_1, γ_t and δ_k are parameters to be estimated, e_t = error term.

The alternative hypothesis is that the series is integrated of order 0, i.e. that no differencing of the series is needed to produce stationarity. Stationary series have a finite variance, transitory innovations from the mean, and a tendency for the series to return to its mean value. In contrast, the non-stationary series is one where the absolute value of the parameters is greater or equal to 1. Non stationary series have a variance which is asymptotically infinite; the series rarely cross the mean (in finite samples), and innovations to the series are permanent [14]. If P_t is found to be non-stationary then it should be determined whether P_t is stationary at first difference, if it is not stationary at first difference, it can also be determined at second difference until stationarity is achieved before we can test for cointegration.

3.3 Test for Cointegration

Having established that the two series are integrated of order 1, cointegration was carried out using Johansen cointegration test. Cointegration implies that there is a linear long-run relationship between price series in spatially separated markets, and is interpreted as a test that $r \neq 0$. This means there must be linear combination (r) of the series. The null hypothesis is that no cointegrating vector existing in the two markets against the alternative of existence of cointegrating vector in the two markets.

Ho: r = 0 (no cointegrating vector existing in the two markets)
Ha: r ≠ 0 (there is existence of cointegrating vector in the two markets)

If $r \neq 0$, it means that the price series are cointegrated and a long-run equilibrium relationship exists between the prices. The Johansen's cointegration method rejects the null

given as follows:

hypothesis of no cointegration (r=0) when the log-likelihood of the unconstrained model that includes the cointegrating equations is significantly different from the log-likelihood of the constrained model that does not include the cointegrating equations. Cointegration tests for market integration is therefore tests of whether there is a statistically linear relationship between different data series [21]. If the two markets are cointegrated, the Johansen trace statistic will be greater than the critical value at both 1% or 5% level of probability. Also, the maximum Eigen statistic will be greater than the critical value at both 1% or 5% level of probability and vice versa. The equation for cointegration in markets i and j is as below:

$$P_{it} = \alpha + \beta_1 P_{jt} + e_t \quad (2)$$

Where P_{it} is price in market i time t, P_{jt} is price in market j time t, α and β_1 are parameters to be estimated, e_t = error term.

If the two markets are cointegrated, we can therefore proceed to Granger causality test to look at the direction of relationship and which market influences the other. Then, the next stage of error correction model (ECM). Cointegration of order 1,1 tests whether there is some linear combination of two series, both of which are integrated of order 1. If a linear combination of the two series is integrated of order 1, then the error term in the linear combination will be integrated of order 0. If the time series are not cointegrated, any linear combination of them will make them non stationary, and so will their residuals be.

4. RESULTS AND DISCUSSION

4.1 Trends in the Prices of Hides in Kano and Lagos Markets (1988 – 2009)

Fig. 1 showed the trends in the average prices of hides in Kano and Lagos. The trends in the prices of hides in both Kano and Lagos markets were very similar. The average prices of hide was ₦500 and ₦550 per one whole piece (2.8kg) of hide between 1988 and 1999 in Kano and Lagos respectively. The average price of hide almost doubled between 1990 and 1993 in both locations. There was about 17% increase in the

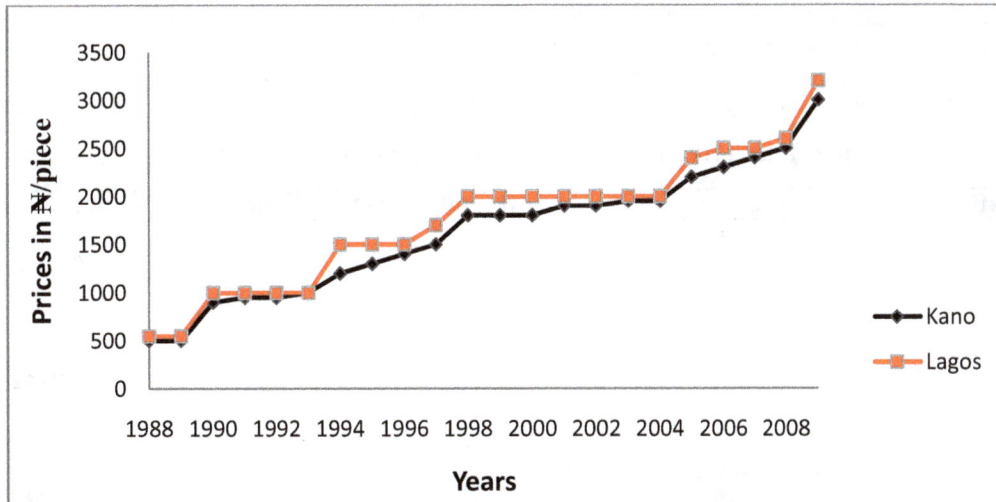

Fig. 1. Trends in the average prices of hides in Kano and Lagos (1988 - 2009)

prices of hide in Kano between 1994 to 1996 (from ₦1200 to ₦1400 respectively) whereas the price was stable in Lagos at ₦1500. The price increased from ₦1500 to ₦1950 in Kano between 1997 and 2003 while it rose from ₦1700 to ₦2000 in Lagos. Between 2004 to 2009, there were similar trends in the price of hides in both locations, though the prices in Kano market were slightly lower than that of Lagos markets in all cases, which may be attributed to cost of transportation of hides from Kano to Lagos before processed to 'pomo'. However, the continuous increase in the prices of hides in both locations over the years may be attributed to inflation and increase in the demand for hides by both the tanners and 'pomo' marketers.

4.2 Unit Root Test Result

The result of the unit root test for prices of hides and 'pomo' in Kano and Lagos markets respectively is presented in Tables 1 and 2. This is a test for stationarity in the two markets. The stationarity test was carried out using Augmented Dickey –Fuller (ADF) test.

In both Kano and Lagos locations, various levels of the Augmented Dickey –Fuller (ADF) test were conducted as follows: no intercept, with intercept, with trend and intercept, with no difference and at first difference. At all these levels of tests, the ADF statistical test values were less than the critical value at 1% and 5% level of probability, indicating that the series were not stationary and need to be differenced. At first difference with intercept, intercept and trend, The ADF statistical values were less than the critical values in

absolute terms at 1% level of probability but greater than the critical values in absolute terms at 5% level of probability in Kano. In Lagos, the ADF value was greater than the critical value at both 1% and 5% levels at first difference. Therefore, the null hypothesis of order 1 was accepted while the alternative was rejected. This implies that the series in the two markets must be differenced once in order to achieve stationarity.

4.3 Cointegration Test Result

Table 3 shows the cointegration results for Kano and Lagos markets. The co-integration test was conducted using Johansen test [22]. The null hypothesis tested was that the two series are co-integrated of order 0, that is, the prices of hides and 'pomo' in the two markets (Kano and Lagos respectively) are not co-integrated. The alternative hypothesis was that the series are co-integrated of order 1, that is, there is co-integration of the series in the two markets.

The result of the Johansen test showed that the Johansen trace statistic (9.459344) was less than the critical value at both 1% and 5% (20.04 and 15.41) level of probability respectively. Since the trace statistic is less than the critical value of at 1% and 5% critical value, it is possible to accept the null hypothesis of no cointegrating vectors and reject the alternative of cointegrating vectors in the two markets. Also, the maximum Eigen statistic (9.311) was less than the critical value at 1% and 5% (18.63 and 14.07) critical value. We can accept the null hypothesis of no co-integrating vectors in the prices of hides and 'pomo' in the two markets.

For the adjustment coefficient in Kano (0.649431), when the standard error (0.55780) is multiplied by 2, it is greater than the adjustment coefficient. For Lagos, the adjustment coefficient is 1.643387, when the standard error (0.74892) is multiplied by two, it is less than the adjustment coefficient. These are also an indication of no cointegration in the two markets.

Table 1. Unit root test result of prices of hides in Kano markets

Location	Levels	No difference*	1st difference*	ADF value
KANO	No intercept			
		-3.8067*** -3.0199** `- 2.6502*		-0.0248
	With intercept			
		-3.7856*** -3.0114** -2.6457*		-0.3742
	With trend and intercept			
		-4.4691*** -3.6454** -3.2602*		-2.1524
	With intercept			
			-3.8304*** -3.0294** -2.6552*	-3.0842
	With trend and intercept			
			-4.5000*** -3.6591** -3.2677*	-4.0776

*** = Significangt at 1% level of probability;** = Significant at 5% level of probability *= Significant at 10% level of probability

Table 2. Unit root test result of prices of 'pomo' in Lagos market

Location	Levels	No difference*	1st difference*	ADF value
LAGOS	No intercept			
		-3.7856*** -3.0114** -2.6502*		-0.0820
	With intercept			
		- 4.4691*** -3.6454** -3.2602*		-2.5225
	With trend and intercept			
		-4.5000*** -3.6591** -3.2677*		-2.6877
	With intercept			
			-3.8304*** -3.0294** -2.6552*	-4.1938
	With trend and intercept			
			-4.5348*** -3.6746** -3.2762*	-3.8321

*** = Significangt at 1% level of probability;** = Significant at 5% level of probability;*= Significant at 10% level of probability

Table 3. Johansen co-integration test result

Hypothesized no of cointegration (s)	Eigen value	Trace statistics	5% critical value	1% critical value
None	0.372215	9.459344	15.41	20.04
At most 1	0.007383	0.148204	3.76	6.55
Hypothesized No of cointegration(s)	Eigen value	Max-eigen statistics	5% critical value	1% critical value
None	0.372215	9.311140	14.07	18.63
At most 1	0.007383	0.148204	3.76	6.65
Adjustment coefficients (standard error in parenthesis)				
D(Kano)	0.649431(0.55780)			
D(Lagos)	1.643387(0.74892)			

However, the reason for no co-integration in the prices of hides in the two locations may be attributed to the fact that there is high competition between the 'pomo' marketers and the tanners in Nigeria. Thus, there was no cointegrating vector between the price of hide in Kano and that of 'pomo' in Lagos.

5. CONCLUSION AND RECOMMENDATIONS

The study found that the series in the two markets were non stationary and need to be differenced once before stationarity is achieved. Also, Johansen cointgration result revealed that the prices of hides and 'pomo' in Kano and Lagos respectively were not cointegrated, eventhough, there was similar trend in prices in the two markets. This implies that there was no long run relationship between the prices of hides in Kano and that of 'pomo' in Lagos The reason for the prices in the two locations having similar trends may be due to exogenous factors like inflation and festive periods like Sallah and Christmas.

Based on the findings, the following recommendation is made. There should be government's commitment to policies that can reduce inflation rate so that prevailing market prices of hides and skins can be stable in different market locations.

COMPETING INTERESTS

Authors have declared that no competing interests exist.

REFERENCES

1. United Nations Devalopment Programmes (UNDP) Leather Production and export in Nigeria. Product Indusry Panel; 2005.

2. Chemonics International Inc Subsector Assessment of the Nigerian Hides and Skins Industry, A paper prepared for The United States Agency for International Development (USAID)/Nigeria Agricultural Development Assistance in Nigeria (ADAN); 2002.

3. Ihuoma AA, Okezie NO, Okonkwo EM, Zubair Y. Current status of the Nigerian leather industry, part 2: Potential hide and skin production. J. Soc. Leather Tech. and Chem. 2001;85(5):170-182

4. Ali A, Rahman SA. Cointegration and market integration: An application to selected sorghum and maize markets in Nigeria, in Olojede AO, Okoye BC, Ekwe KC, Chukwu GO, Nwachukwu IN, Alawode O. Editors Proceedings of the Annual Conference of the Agricultural Society of Nigeria (ASN); 2009, held in University of Abuja; 2009.

5. Trotter BW. Applying price analysing to marketing systems: Methods and examples from the indonesian rice market. Marketing Series Natural Resources Institute, Chatham, U.K. 1992;3.

6. Granger CWJ. Some properties of time series data and their use in econometric model specification. Journal of Econometrics; 1981.

7. Lence S, Falk B. Cointegration, market integration and market efficiency. J. Int'l Money Fin. 2005;24:873-890.

8. Rahman SA. Techniques for analyzing market integration in the agricultural sector. Paper presented at the Nigeria Association of Agricultural Economist (NAAE) Annual conference. Bello University, Zaria, Nigeria; 2004.

9. Dickey DA, Fuller WA. Distribution of the estimators for autoregressive time series with a unit root. Journal of American Statistical Association. 1979;74:427-431.

10. Dickey DA, Fuller WA. The likelihood ratio statistics for autoregressive time series with a unit root. Econometrica. 1981;49:1057–1072.

11. Evans GBA, Savin NE. Testing for unit root: 1, Econometrica. 1981;49:753-779.

12. Granger CWJ, Weiss AA. Time series analysis of error – Correcting models, in studies in econometrics, time series, and multivariate statistics, New York: Academic Press. 1983;255- 278.

13. Jaleta M, Gebermadhin B. Price cointegration analysis of food crop markets: The case of wheat and Teff Commodities in Northern Ethiopia, Contributed Paper Prepared for Presentation at the International Association of Agricultural Economists Conference, Beijing, China; 2009.

14. Akande SO, Akpokodje G. The Nigerian rice economy in a competitive world: Constraints, opportunities and strategic choices, rice prices and market integration in selected areas in nigeria, agriculture and rural development department, Nigerian Institute of Social and Economic Research (NISER), Ibadan; 2003.

15. Oyakhilomen O, Omadachi UO, Zibah RG. Cocoa production – agricultural credit guarantee scheme fund Nexus in Nigeria: A cointegration approach. Russian J. Ag. Socioecon. Sci. 2012;9(9):28–32.

16. Aminu A. Marketing of tomato and onion in Jigawa State: A spatial and seasonal price analysis, Ph. D. Dissertation, Department of Agricultural Economics and Rural Sociology, Ahmadu Bello University, Zaria; 2004.

17. Mukhtar T, Javed MT. Market integration in maize markets in Pakistan. Regional and Sectoral Economic Studies. 2008;8(2):85-96.

18. Chirwa WE. Food marketing reforms and integration of maize and rice markets in Malawi. Working Papers No. WC/05/99; 1999.

19. Engle RF, Granger CWJ. Cointegration and error correction: Representation, estimation and testing. Econometrica. 1987;55(2):251–276.

20. Nigerian Tanners Council, List of tanners and their addresses, Tanners council registered in Nigeria, 31/32 Niger Street, Kano; 2009.

21. Asche F, Bremnes H, Wessells CR. Product aggregation, market integration, and relationships between prices: An application to world salmon markets. Amer. J. Ag. Econs; 1999;81:568-581.

22. Johansen S. Estimation and hypothesis test of cointegration vectors in Gaussian vector autoregressive models, Econometrica. 1991;59(6):1551-1580.

22

Prevalence of Endometritis and Its Associated Risk Factors in Dairy Cattle of Central Uganda

Dickson Stuart Tayebwa[1*], Godfrey Bigirwa[1], Joseph Byaruhanga[1] and Keneth Iceland Kasozi[1,2]

[1]Central Diagnostic Laboratory, Department of Veterinary Pharmacy Clinics and Comparative Medicine, College of Veterinary Medicine Animal Resources and Biosecurity, Makerere University, Kampala, Uganda.
[2]Kiboga Regional Veterinary Laboratory, Department of Production and Marketing, Kiboga Local Government, Kiboga, Uganda.

Authors' contributions

This work was carried out in collaboration between all authors. Authors DST, GB designed the study and wrote the study protocol and collected the data. Authors DST, JB and KIK wrote the first draft of the manuscript, managed analysis and performed statistical analysis. All authors read and approved the final manuscript for submission.

Editor(s):
(1) Rusu Teodor, Department of Technical and Soil Sciences, University of Agricultural Sciences and Veterinary Medicine Cluj-Napoca, Romania.
(2) Anonymous.
Reviewers:
(1) Anonymous, Venezuela.
(2) Anonymous, Kosovo.
(3) Juan José Romero Zúñiga, Population Medicine Research Program, National University, Costa Rica.
(4) Anonymous, India.

ABSTRACT

Aims: Endometritis is a major postpartum disease that affects dairy cattle productivity which is accompanied by heavy economic losses to the farmer. The status of Clinical endometritis (CE) and Sub-Clinical Endometritis (SCE) in sub-Saharan Africa is poorly understood, thus the study was carried out to provide information on the prevalence and associated risk factors that are responsible for the occurrence of SCE and CE in dairy cattle of Central Uganda.
Study Design: This was a prospective study involving 140 dairy cattle within 60 days postpartum from 35 commercial dairy farms in Central Uganda. The dairy herds were visited in both the dry (n=90) and wet season (n=50) and subsequent visits were conducted after 3 months and 5 months

Corresponding author: E-mail: dtayebwa@covab.mak.ac.ug;

to collect data for reproductive performance

Methodology: A metricheck® device was used to collect the cervico-vaginal discharge which was examined for color, odor, texture, and measurement of its pH during the postpartum period for diagnosis of endometritis. Further examination of the reproductive tract was carried out using a vaginal speculum and subsequently rectal palpations were performed. The objectives of the study were to determine the prevalence of CE and SCE in the Central Uganda and assess the risk factors involved.

Results: In this study, the prevalence of CE and SCE was established at 3.6% and 18.6% respectively, this burden was slightly higher in the wet than in the dry season but with no statistical significance P=0.126. Dairy cattle that had calved more than three times were shown to be associated with a higher body appearance (BCS > 3) than those that had calved down fewer times. Dystocia, Retained After Birth and Abortion were identified as associated risk factors (P = 0.00) to SCE and CE whereas infertility and Prolonged days calving to conception (>90 d) were postpartum implications (P=0.00) associated with SCE and CE in this study. This would be attributed to the poor management of postpartum dairy cattle in the farms visited as no farm was found with a maternity pen. SCE caused infertility in 65.5% of the dairy cattle whereas the CE is a major influencing factor to long calving to first AI and calving to conception interval (306d±90.6), On basis of reproduction, there was no major difference towards use of AI or Natural service.

Conclusion: Management of endometritis in the region should address pre-partum and postpartum dairy herd management through improved extension service delivery and technical farm support to construct maternity pens, Endometritis is a multifactorial disease that requires a multidisciplinary approach to boost nutrition and health thus reducing incidence of risk factors such as dystocia and Retained fetal birth) .subsequent studies should be carried out to explore the national burden of SCE and ascertain the cause of the abortion.

Keywords: Endometritis in dairy cattle; dairy production systems in Uganda; reproductive performance.

1. INTRODUCTION

Microbial disease of the female reproductive tract is a major concern in dairy cattle production systems [1]. This is because microbial infections are highly associated with infertility as they disrupt uterine and ovarian function. Uterine disease within a week of parturition (metritis) has been shown to be common to dairy cattle [2]. The clinical presentation of endometritis has been clearly elaborated in previous studies, and it has been shown to include both Clinical Endometritis (CE) and Sub-Clinical Endometritis (SCE) [2-5]. CE has been shown to be characterized by an enlarged uterus and a watery red-brown fluid to viscous off-white purulent uterine discharge, which often has a fetid odor [1]. SCE on the other hand is associated with inflammation of the endometrium that results in a significant reduction in reproductive performance in the absence of signs of clinical endometritis [6]. The inflammation is presumably associated with recovery of the tissues after clinical endometritis, trauma or other non-microbial disease. SCE has been shown to be defined by polymorphonuclear neutrophils (PMNs) exceeding between 5.5% of cells and 10% of cells in samples collected by flushing the uterine lumen [1].

The prevalence of SCE and CE in Sub-Saharan Africa is not clearly known, but studies in Germany have shown it to be close to 15% to 20% in dairy cattle respectively [7,8]. In dairy cattle, uterine infections are not life threatening and often unavoidable; however, they reduce fertility and increase the production costs of properties [9]. The major risk factors that have been identified associated with the endometritis include: health status, parity and body condition score (BCS) of cows, and calving date and it has been shown that the first service conception rate is lower in dairy cattle suffering from endometritis which significantly decreases reproductive performance of dairy herds [10]. The study was carried out to provide information on the prevalence and assessment of risk factors that are responsible for the occurrence of sub/clinical endometritis in Central Uganda.

2. MATERIALS AND METHODS

2.1 Farms and Animals

This was a prospective study carried out for a period of 13 months involving dairy farmers in the Central region (Kampala, Mukono and Wakiso districts) of Uganda. A total of 35 Commercial dairy farms were randomly visited for examination of dairy cows (n=140) within 60 days postpartum, involving both local and exotic dairy breeds. Herd size was not an inclusion criteria but the farms visited were free stall (45.7%) majority, paddocking (31.4%), Zero grazing (22.9%) with average herd size of 35.8±0.05, 70.3±22.5 and 12±10.1 respectively. Classification of the management system was based on; farms keeping their animals strictly in a pen feeding them on cut and carry grass, forage with peelings from bananas (*Musa acuminata*) supplements as zero grazing farms, those with a stall for feeding the cows on cut and carry grass, forage but with one or two paddocks around the stall where the animals graze, free stall; while paddocking farms were those that entirely let the animals graze on pasture within fenced perimeters without any supplementation. The farmers were visited initially once in the dry (n=90, June - August 2013) and wet (n=50, April - June) seasons in a study area that experiences a bimodal rainfall pattern through the year separated by two short dry seasons from December to March and June to August. Monitoring visits (2), after 3 and later at 5 Months were scheduled for monitoring reproductive performance (return to heat, service and conception). All dairy cattle within 60 days postpartum were included in the study using farm reproductive records, among details taken were: breed, age, date of last calving, date of last AI (Artificial Insemination), calf health, prior to the physical examination and rectal palpation. The stage of parity was classified into three groups i.e. young (< 2 times), prime (3 times), mature (> 3 times). The body condition score (BCS) was recorded using a rank of 1-5 [11]; the dairy production systems and farm descriptive data were recorded through observations that were made. Any complications at the time of calving were recorded by use of farm records and discussion with the farmer or herdsman and those taken into account included, dystocia; if the cow had assisted delivery, Abortion; when the cow lost the fetus before term, Retained After birth; if expulsion of the placenta was assisted after 24hrs and non-reproductive related diseases such as tick borne diseases were classified as ill health.

2.2 Sample Collection and Reproductive Examination

A metricheck® device was used to collect the cervico-vaginal discharge and observations for the characteristic of the discharge (appearance, color, texture odor) were made and recorded, the pH of the discharge was measured using a pH test paper (BTB role type, Advantec®). Subsequently, further examination was carried out using a vaginal speculum (4 cm in diameter and 35 cm in length) and a torch light for observations of the cervix status (closed or open), mucous and or pus origin (uterus or vagina) and rectal palpation was performed to check the status of the ovaries, cervix and uterine horns as well as any other anomalies in the reproductive system [12,13], Endometritis was clinically diagnosed based on discharge characteristics and pH reading, the discharge was considered normal if it was clear or slightly cloudy without pus specks, a foul smell and pH (6.4-6.8). Mucopurulent (approximately 50% pus and 50% mucous with no foul smell and pH > 7.0) was considered SCE and purulent (> 50% pus with foul smell, pH > 7.2) was considered abnormal, indicating CE [8]. Microscopic examination of the cervico-vaginal discharge was not conducted.

Reproductive performance data was collected after 3 and 5 months, records were taken for; date when seen on heat, date of service (AI/bull), production disease (Milk fever, Downers cow syndrome, Mastitis), rectal palpation was done to confirm pregnancy of animals served and check for cases of Infertility (cystic ovaries, silent heat, acyclic ovaries, return to heat), animals that had not conceived were subjected to further examination using a vaginal speculum to check for health status of the reproductive tract.

2.3 Statistical Analysis

Data was recorded in Microsoft excel (version 2010) before transferring it to SPSS statistics data editor (version 17) for expression of frequencies and descriptive data using Chi-square at a Confidence Interval (CI 95%), a P value < 0.05 was considered statistically significant for all the tests performed. The graphical output was done with (Graph Prism version 6).

3. RESULTS AND DISCUSSION

The study revealed that the prevalence of Clinical Endometritis (CE) and Sub-clinical Endometritis (SCE) is at 3.6% and 18.6% in the study as shown in Table 1.

It was shown that the prevalence of Endometritis was the slightly higher in the rainy season than in the dry season but no statistically significant differences existed (P=0.126) regarding the prevalence. The predominant breed in the study was the Holstein Friesian thus the mostly affected by both CE and SCE, Following the government's recent efforts to improve the livestock industry through the National Agricultural Advisory Directorate Services (NAADs) project that provided mainly Holstein Friesian to farmers countrywide, the major breeds of most farmers in Central Uganda were found to be composed of Bos taurus (Table 2). The majority of the dairy cattle affected by SCE were shown to be associated with a BCS > 3, as well as those with parity ≥ 2 as shown in Table 2.

Further analysis of the different breeds and location showed no marked inferences (P > 0.05). Dystocia, abortion and retained after birth were identified as the major risk factors to both CE and SCE in this study (P=0.003), while infertility was shown to be the major postpartum complication in dairy cattle that suffer from SCE (69.2%, P=0.000). the study showed that 100% of cattle that suffered from CE had a long calving to conception interval >180 days, this is scenario is quite replicated SCE, where 76.9% of animals that suffered this disease had days calving-conception > 180 (P=0.000), this hinders achieving the reproductive index of a calf per cow per year, thus imposing loss to the dairy farmer.

The SCE prevalence reported in this study highlights the risk raised by several sub-clinical reproductive infections in the dairy industry as it has been shown to have high economic implications [7]. This is important since farmers in the region have often been reported to institute treatment without necessarily seeking the services of professional veterinarian as a result of the liberal drug industry [14,15]. This has continuously curtailed the farmer and animal health scientists to easily diagnose diseases that continue to cripple the productivity of the livestock industry especially those of sub clinical nature. The identified risk factors; RAB, abortion and dystocia are thought to be responsible for the poor involution of the uterus thus causing the high prevalence of SCE, thus leading to infertility and lengthened calving-conception which is in agreement with a recent study [1]. The incidence of endometritis i.e. both CE and SCE was shown to be relatively the same in both seasons, illustrating that disease occurrence is independent of the time of study and not associated with Endometritis (P = 0.126) which was in agreement with recent findings [1].

The risk factors; dystocia and RAB were strongly associated with occurrence of SCE and CE in this study. This would be due to the poor management, no farm had a maternity pen and feed supplementation program for cows in dry period among the farms visited. Dairy cattle that had a higher body condition score (BCS > 3) were associated with a higher burden of SCE contrary to what had been illustrated in a previous study [1]. These discrepancies may reflect differences in the geography, the study population and management practices. The drop in BCS from the previous parturition postpartum has been shown to increase pregnancy loss probably accounting for the low parity associated with this group [9]. Co-existing nutritional deficiencies that the postpartum dairy cattle probably would be suffering from such as Sub-Clinical Ketosis (SCK) are probably responsible for the loss of the protective advantage gained prepartum [16]. The abnormally lengthened calving intervals due to endometritis have been shown to be associated with low productivity and thus serious financial implications to the farmer as productivity of the livestock herd is reduced severely [5,17]. This is partly responsible for the poor reproductive performance of most of the dairy cattle in Central region thus making the realization of the millennial developmental goals impossible at the moment. Like in many cases, endometritis as shown in Table 2, the most prevalent is the subclinical form (SCE, 18.6%) thus treatment and management is very minimal because farmers and extension workers have limited knowledge to diagnose this disease, [15]. The bacteria often infect the endometrium and proliferate without ceasing causing high cases of infertility. A recent study has shown that the accumulation of interleukins and ovarian steroids estradiol and progesterone have little or no impact on inflammatory responses to inflammation of the endometrium [2], as such proper diagnosis, management and treatment with antibiotics is the way to reduce the impact of this disease and improve productivity and income of the dairy farmer in Central Uganda.

This study highlighted that there is significant relationship between the burden of SCE and infertiity rate as illustrated in Table 3 with the highest rate 65.5 recorded for those animals that suffered from SCE (P=0.000, CI 95%), whereas CE has been shown to highly increase the calving to conception interval to over a year. This probably explains the extremely low conception rate shown at first service for animals suffering from both CE and SCE. This does not pinpoint but illustrates the potential of CE and SCE to infertility, culling and loss to the farmer.

Incidentally there was no significant difference in the success of AI or natural service, similary with the conception rate.

The study showed a peculiar trend between the calving's and the rainfall pattern (seasonal variations) with the highest number of calving's (41.4%) registered in July during the dry season, the month with the least amount of rainfall registered per year as shown in Fig. 1. This may explain the low burden of CE in regards to the reduced chances of contamination at the time of parturition given the management state of the farms in the region that have no maternity pens, cows tend to produce from the pens or the farms and if this coincides with high rainfall and a damp, dirty environment, complimented by farmers self-reliance to assist parturition without seeking veterinary help. Such natural circumstances like cows calving in the dry season may contribute positively to the low prevalence of CE and SCE.

Table 1. Showing the overall prevalence of endometritis in commercial dairy cattle in Central Uganda

Endometritis	Clinical endometritis / CE (%)	Sub-clinical endometritis / SCE (%)	Negative	Total
Prevalence	5 (3.6)	26 (18.6)	109 (77.8)	140 (100)

Table 2. Showing CE and SCE prevalence and factors associated with its occurrence in dairy cattle

PARAMETERS	VARIABLES	Normal	CE	SCE	Total	Chi square & P- value, CI (95%)
Breed	Holstein Friesian	93(85.3%)	4(80.0%)	22(84.6%)	119(85%)	
	Guernsey	10(9.2%)	0(0%)	1(3.8%)	11(7.9%)	
	Jersey	2(1.8%)	1(20.0%)	2(7.7%)	5(3.6)	7.305
	Indigenous	4(3.7%)	0(0%)	1(3.8%)	5(3.6%)	0.294
Body Condition Score	< 3	26(23.9%)	2(40.0%)	7(26.9%)	35(25%)	
	= 3	25(22.9%)	1(20.0%)	4(15.4%)	30(21.4%)	1.366
	>3	58(53.2%)	2(40.0%)	15(57.7%)	75(53.6%)	0.850
Parity	< 2	47(43.1%)	0(0%)	10(38.5%)	57(40.7%)	3.75
	≥ 2	62(56.9%)	5(100%)	16(61.5%)	83(59.3%)	0.153
Season	Dry	74(67.9%)	2(40.0%)	13(50.0%)	89(63.6%)	4.145
	Wet	35(32.1%)	3(60.0%)	13(50.0%)	51(36.4%)	0.126
Location	Kampala	18(16.5%)	1(20.0%)	2(7.6%)	21(15%)	
	Wakiso	30(27.5%)	2(40.0%)	12(46.2%)	44(31.4%)	4.171
	Mukono	61(56.0%)	2(40.0%)	12(46.2%)	75(53.6%)	0.383
Risk factors						
Complications at Birth	Non	103(94.5%)	0(0%)	19(73.1%)	122(87.1)	
	RAB	0(0%)	2(40.0%)	1(3.8%)	3(2.1%)	
	Dystocia	0(0%)	1(20.0%)	2(7.7%)	3(2.1%)	
	Abortion	0(0%)	2(40.0%)	1(3.8%)	3(2.1%)	92.915
	Ill health	6(5.5%)	0(0%)	3(11.5%)	9(6.4%)	0.000
Postpartum implication						
Calving to conception	Within 90 Days	27(24.8%)	0(0%)	2(7.7%)	29(20.7%)	
	90-180 Days	37(33.9%)	0(0%)	4(15.4%)	41(29.3%)	15.943
	> 180 Days	45(41.3%)	5(100%)	20(76.9%)	70(50%)	0.003
Postpartum Complications	Non	90(82.6%)	4(80.0%)	4(15.4%)	98(70%)	
	Infertility	5(4.6%)	1(20.0%)	18(69.2%)	24(17.1%)	
	Production Diseases	14(12.8%)	0(0%)	4(15.4%)	18(12.9%)	65.579 0.000

Key: CE = Clinical endometritis; SCE = Sub-Clinical endometritis; RAB=Retained after Birth

Table 3. Reproductive performance of dairy cattle in Central Uganda

	Normal	CE	SCE	Total
		Animal health status		Total
Number of animals	109	5	26	140
Conception rate at first AI service (%)	31.2	20.0	3.8	25.7
Conception rate at first service by Bull (%)	33.9	40.0	11.5	30.0
Conception rate at ≥ 2 AI services (%)	2.8	-	7.7	3.6
Conception rate at ≥ 2 services by Bull (%)	4.6	-	3.8	4.3
Infertility rate (%)	12.8	20.0	65.5	22.8
Culling rate (%)	14.7	20.0	7.7	13.6
Calving to first AI interval (mean ± SD)	133.3±82.5	306±90.6	137.6±91.5	139.4±88.1
Calving to conception interval (mean ± SD)	138.7±79.5	306±90.6	143.3±88.3	144.7±85.1

Key:CE; Clinical Endometritis, SCE; Subclinical Endometritis, AI; Artificial Insemination

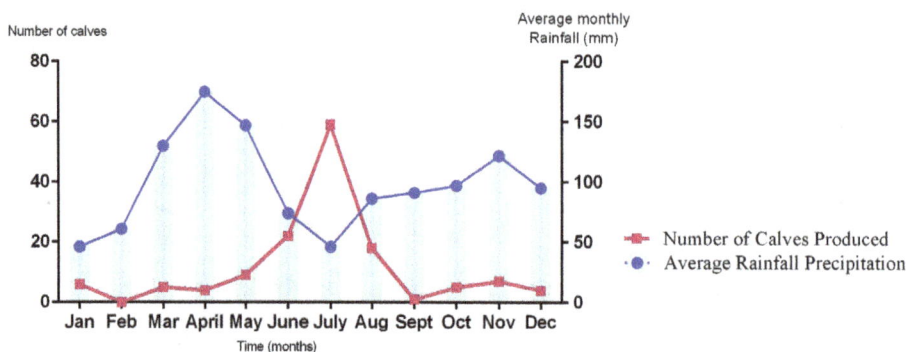

Fig. 1. Distribution of the calvings per month in relation to rainfall pattern (world weather online/Kampala-weather-averages, data set from 2000-2012)

4. CONCLUSION

The prevalence of CE and SCE was established at 3.6% and 18.6% respectively and the major risk factors associated with the high SCE burden were dystocia, RAB and abortion. Having more than 2 calves was also associated with a high disease burden, which is thought to be due to the management practices of the herds in the study area. Endometritis status observed in this study is reflects on the need to for farmers support bodies like NAADS and extension service providers like to concentrate on providing knowledge on management and awareness of production constraining diseases with a subclinical nature such as Endometritis and among others Mastitis. The liberal nature of the drug Industry in Uganda has cultivated a habit of self-reliance of farmers at handling disease without help from professional because they can access any type of drug over the counter and care much less about prescription and proper application in terms of dosage and dosage rate, this has elevated the prevalence of such disease as in discussion because the farmers attempt to assist delivery thus causing dystocia and contamination majority of the times thus the high prevalence of SCE.

A multifactorial disease like Endometritis needs a multidisciplinary approach, thus farmers should be sensitized on proper feeding with high roughage feed (hay), corn (Zea mays) silage and supplementation with mineral and salt leaks would help to boost immunity and reduce cases of dystocia and RAB contrary to feeding animals with Elephant / Napier grass (Pennisetum purpureum) and peelings from bananas (Musa acuminata). Subsequent studies should be carried out to explore fully major limiting factors in relation to SCE in the study area. Abortion was among strongly associated with burden of CE and SCE, all the animals that aborted suffered from either of the two yet no data is substantial data is available on the causes of brucellosis in Uganda over the past seven (7) Years.

This study shows that there is no significant difference in the success when one uses AI or natural service, however it is highly suggested that more farmers adopt the use of AI for its economic and health soundness.

ACKNOWLEDGEMENTS

The authors would like to appreciate the Contribution of the staff of the Central Diagnostic Laboratory, the farmers; who accepted for their animals to be tested and the invaluable support of Mr. Godfrey Luwemba; the extension worker and artificial Inseminator of the College of Veterinary Medicine, Animal Resources and Biosecurity, Makerere University. This work was carried out as part of the Japanese Technical Assistance to Improve Animal Disease Diagnostics and partially funded by Japan International Cooperation Agency (JICA) in Uganda.

COMPETING INTERESTS

Authors have declared that no competing interests exist.

REFERENCES

1. Kasimanickam RK, Kasimanickam VR, Olsen JR, Jeffress EJ, Moore D, Kastelic JP. Associations among serum pro- and anti-inflammatory cytokines, metabolic mediators, body condition, and uterine disease in postpartum dairy cows. Reproductive Biology and Endocrinology: RB&E. 2013;11:103. DOI:10.1186/1477-7827-11-103.

2. Turner ML, Cronin JG, Healey GD, Sheldon IM. Epithelial and stromal cells of bovine endometrium have roles in innate immunity and initiate inflammatory responses to bacterial lipopeptides in vitro via Toll-like receptors TLR2, TLR1, and TLR6. Endocrinology. 2014;155(4):1453-65. DOI:10.1210/en.2013-1822.

3. Wathes DC, Cheng Z, Fenwick MA, Fitzpatrick R, Patton J. Influence of energy balance on teh somatotrophic axix and matrix metalloproteinase expression in the postpartum dairy cow. Reproduction. 2011;141:269-281. DOI:10.1530/REP-10-0177.

4. Patra MK, Kumar H, Nandi S. Neutrophil functions and cytokines expression profile in buffaloes with impending postpartum reproductive disorders. Asian-Australasian Journal of Animal Sciences. 2013;26(10):1406-15. DOI:10.5713/ajas.2012.12703.

5. Strüve K, Herzog K, Magata F, Piechotta M, Shirasuna K, Miyamoto A. The effect of metritis on luteal function in dairy cows. BMC Veterinary Research. 2013;9(244):1-9.

6. Herath S, Lilly ST, Santos NR, Gilbert RO, Goetze L, Bryant CE, Sheldon IM. Expression of genes associated with immunity in the endometrium of cattle with disparate postpartum uterine disease and fertility. Reproductive Biology and Endocrinology: RB&E. 2009;7:55. DOI:10.1186/1477-7827-7-55.

7. Kaufmann TB, Drillich M, Tenhagen BA, Heuwieser W. Correlations between periparturient serum concentrations of non-esterified fatty acids, beta-hydroxybutyric acid, bilirubin, and urea and the occurrence of clinical and subclinical postpartum bovine endometritis. BMC Veterinary Research. 2010;6(1):47. DOI:10.1186/1746-6148-6-47.

8. Gabler C, Fischer C, Drillich M, Einspanier R, Heuwieser W. Time-dependent mRNA expression of selected pro-inflammatory factors in the endometrium of primiparous cows postpartum. Reproductive Biology and Endocrinology. 2010;8(1):152. DOI:10.1186/1477-7827-8-152.

9. Carneiro LC, Ferreira AF, Padua M, Saut JP, Ferraudo AS, Dos Santos RM. Incidence of subclinical endometritis and its effects on reproductive performance of crossbred dairy cows. Tropical Animal Health and Production. 2014;46(8):1435-9. DOI:10.1007/s11250-014-0661-y.

10. Kim IH, Kang HG. Risk factors for postpartum endometritis and the effect of endometritis on reproductive performance in dairy cows in Korea. The Journal of Reproduction and Development. 2003;49(6), 485-491.

11. Elanco. Body Condition Scoring in Dairy Cattle. 2009;1-8.

12. McDougall S, Macaulay R, Compton C. Association between endometritis diagnosis using a novel intravaginal device and reproductive performance in dairy cattle. Animal Reproduction Science. 2007;99(1-2):9-23.

13. Lambertz C, Völker D, Janowitz U, Gauly M. Evaluation of vaginal discharge with the Metricheck device and the relationship to reproductive performance in postpartum dairy cows. Animal Science Journal Nihon Chikusan Gakkaihō. 2014;85(9):848-52. DOI:10.1111/asj.12219.

14. Florence Kasirye NM, Ivan T. Dairy Platform: Policy Brief 3. In Management of Livestock drug Systems. 1989;1–3.

15. Kasozi KI, Matovu E, Tayebwa DS, Natuhwera J, Mugezi I, Mahero M. Epidemiology of increasing hemo-parasite burden in Ugandan Cattle. Open Journal of Veterinary Medicine. 2014;4(10):220–231.

16. Knop R, Cernescu H. Effects of negative energy balance on reproduction in dairy cows. Lucrari Stintifice Medicina Veterinara. 2009;42(119):198–205.

17. Lee J, Kim I. Pregnancy loss in dairy cows: The contributing factors, the effects on reproductive performance and the economic impact. J Vet Sci. 2007;8(3):283–288.

Permissions

The contributors of this book come from diverse backgrounds, making this book a truly international effort. This book will bring forth new frontiers with its revolutionizing research information and detailed analysis of the nascent developments around the world.

We would like to thank all the contributing authors for lending their expertise to make the book truly unique. They have played a crucial role in the development of this book. Without their invaluable contributions this book wouldn't have been possible. They have made vital efforts to compile up to date information on the varied aspects of this subject to make this book a valuable addition to the collection of many professionals and students.

This book was conceptualized with the vision of imparting up-to-date information and advanced data in this field. To ensure the same, a matchless editorial board was set up. Every individual on the board went through rigorous rounds of assessment to prove their worth. After which they invested a large part of their time researching and compiling the most relevant data for our readers.

The editorial board has been involved in producing this book since its inception. They have spent rigorous hours researching and exploring the diverse topics which have resulted in the successful publishing of this book. They have passed on their knowledge of decades through this book. To expedite this challenging task, the publisher supported the team at every step. A small team of assistant editors was also appointed to further simplify the editing procedure and attain best results for the readers.

Apart from the editorial board, the designing team has also invested a significant amount of their time in understanding the subject and creating the most relevant covers. They scrutinized every image to scout for the most suitable representation of the subject and create an appropriate cover for the book.

The publishing team has been an ardent support to the editorial, designing and production team. Their endless efforts to recruit the best for this project, has resulted in the accomplishment of this book. They are a veteran in the field of academics and their pool of knowledge is as vast as their experience in printing. Their expertise and guidance has proved useful at every step. Their uncompromising quality standards have made this book an exceptional effort. Their encouragement from time to time has been an inspiration for everyone.

The publisher and the editorial board hope that this book will prove to be a valuable piece of knowledge for researchers, students, practitioners and scholars across the globe.

List of Contributors

Gideon Danso-Abbeam
Department of Agricultural and Resource Economics, University for Development Studies, P.O.Box TL 1882, Tamale, Ghana

Abubakari M. Dahamani
Department of Agricultural and Resource Economics, University for Development Studies, P.O.Box TL 1882, Tamale, Ghana

Gbanha A-S Bawa
Department of Agricultural and Resource Economics, University for Development Studies, P.O.Box TL 1882, Tamale, Ghana

O. A. Adebiyi
Department of Animal Science, Faculty of Agriculture and Forestry, University of Ibadan, Ibadan, Oyo State, Nigeria

U. V. Okolie-Alfred
Department of Animal Science, Faculty of Agriculture and Forestry, University of Ibadan, Ibadan, Oyo State, Nigeria

C. Godstime
Department of Animal Science, Faculty of Agriculture and Forestry, University of Ibadan, Ibadan, Oyo State, Nigeria

O. A. Adeniji
Department of Animal Science, Faculty of Agriculture and Forestry, University of Ibadan, Ibadan, Oyo State, Nigeria

P. K. Ajuogu
Department of Animal Science and Fisheries, Faculty of Agriculture, University of Port Harcourt, P.M.B.5323 Choba, Port Harcourt, Nigeria

U. Herbert
Department of Animal Science, Micheal Okpara Federal University of Agriculture Umudike, Abia State, Nigeria

M. A. Yahaya
Department of Animal Science, Faculty of Agriculture, Rivers State University of Science and Technology, Nkpolu Orowurukwu Diobu, Port Harcourt, Rivers State, Nigeria

I. A. Adebayo
Department of Forestry, Wildlife and Fisheries Management, Faculty of Agricultural Sciences, Ekiti State University, Ado Ekiti, Ekiti State, Nigeria

G. H. Akinwumi
Department of Forestry, Wildlife and Fisheries Management, Faculty of Agricultural Sciences, Ekiti State University, Ado Ekiti, Ekiti State, Nigeria

A. D. Oduro-Owusu
Department of Animal Science Education, University of Education, Winneba, Box 40, Mampong- Ashanti, Ghana

J. K. Kagya-Agyemang
Department of Animal Science Education, University of Education, Winneba, Box 40, Mampong- Ashanti, Ghana

S. Y. Annor
Department of Animal Science Education, University of Education, Winneba, Box 40, Mampong- Ashanti, Ghana

F. R. K. Bonsu
Department of Animal Science Education, University of Education, Winneba, Box 40, Mampong- Ashanti, Ghana

Mulugeta Atnaf
Ethiopian Institute of Agricultural Research, Pawe research center, P.O.Box 25, Pawe, Ethiopia
Department of Microbial, Cellular and Molecular Biology, Collage of Natural Sciences,
Addis Ababa University, P.O.Box 1176, Addis Ababa, Ethiopia

Kassahun Tesfaye
Department of Microbial, Cellular and Molecular Biology, Collage of Natural Sciences,
Addis Ababa University, P.O.Box 1176, Addis Ababa, Ethiopia

Kifle Dagne
Department of Microbial, Cellular and Molecular Biology, Collage of Natural Sciences,
Addis Ababa University, P.O.Box 1176, Addis Ababa, Ethiopia

Aniekan Jim Akpaeti
Department of Agricultural Economics and Extension, Akwa Ibom State University, Ikot Akpaden, Mkpat Enin, P.M.B.1167, Uyo, Akwa Ibom State, Nigeria

L. A. F. Akinola
Department of Animal Science and Fisheries, Faculty of Agriculture, University of Port Harcourt, P.M.B. 5323, Port Harcourt, Rivers State, Nigeria

O. A. Ekine
Department of Animal Science and Fisheries, Faculty of Agriculture, University of Port Harcourt, P.M.B. 5323, Port Harcourt, Rivers State, Nigeria

C. C. Emedo
Department of Animal Science and Fisheries, Faculty of Agriculture, University of Port Harcourt, P.M.B. 5323, Port Harcourt, Rivers State, Nigeria

Joseph S. Ekpo
Department of Animal Science, Akwa Ibom State University, P.M.B.1167, Uyo, Akwa Ibom State, Nigeria

Nseabasi N. Etim
Department of Animal Science, Akwa Ibom State University, P.M.B.1167, Uyo, Akwa Ibom State, Nigeria

Glory D. Eyo
Department of Animal Science, Akwa Ibom State University, P.M.B.1167, Uyo, Akwa Ibom State, Nigeria

Edem E. A. Offiong
Department of Animal Science, Akwa Ibom State University, P.M.B.1167, Uyo, Akwa Ibom State, Nigeria

Metiabasi D. Udo
Department of Animal Science, Akwa Ibom State University, P.M.B.1167, Uyo, Akwa Ibom State, Nigeria

L. Bashir
Department of Biochemistry, Federal University of Technology, PMB 65, Minna, Niger State, Nigeria

P. C. Ossai
Department of Biochemistry, Federal University of Technology, PMB 65, Minna, Niger State, Nigeria

O. K. Shittu
Department of Biochemistry, Federal University of Technology, PMB 65, Minna, Niger State, Nigeria

A. N. Abubakar
Department of Biochemistry, Federal University of Technology, PMB 65, Minna, Niger State, Nigeria

T. Caleb
Department of Biochemistry, Federal University of Technology, PMB 65, Minna, Niger State, Nigeria

S. A. Adesoji
Department of Agricultural Extension and Rural Development, Obafemi Awolowo University, Ile Ife, Osun State, Nigeria

S. I. Ogunjimi
Department of Agricultural Economics and Extension, Federal University, Oye-Ekiti, Ekiti State, Nigeria

S. A. Okunsebor
Department of Agriculture, Shabu-Lafia Campus, Nasarawa State University, Keffi, P.M.B. 135, Lafia, Nasarawa State, Nigeria

K. G. Shima
Department of Agriculture, Shabu-Lafia Campus, Nasarawa State University, Keffi, P.M.B. 135, Lafia, Nasarawa State, Nigeria

T. F. Sunnuvu
School of Agriculture, Lagos State Polytechniques, Ikorodu Lagos, Nigeria

Stephen Opoku–Mensah
Department of Agropreneurship, Institute of Entrepreneurship and Enterprise Development, Kumasi Polytechnic, Kumasi, Ghana

Hayford Agbekpornu
Mininistry of Fisheries and Aquaculture Development, Accra, Ghana

B. B. Okafor
Department of Agricultural Science, Ignatius Ajuru University of Education, Ndele Campus, P.M.B. 5047, Port Harcourt, Nigeria

G. A. Kalio
Department of Agricultural Science, Ignatius Ajuru University of Education, Ndele Campus, P.M.B. 5047, Port Harcourt, Nigeria

H. A. Manilla
Department of Agricultural Science, Ignatius Ajuru University of Education, Ndele Campus, P.M.B. 5047, Port Harcourt, Nigeria

O. N. Wariboko
Department of Agricultural Science, Ignatius Ajuru University of Education, Ndele Campus, P.M.B. 5047, Port Harcourt, Nigeria

A. E. Ede
Department of Crop Science, Faculty of Agriculture, University of Nigeria, Nsukka, Nigeria

U. M. Ndubuaku
Department of Crop Science, Faculty of Agriculture, University of Nigeria, Nsukka, Nigeria

K. P. Baiyeri
Department of Crop Science, Faculty of Agriculture, University of Nigeria, Nsukka, Nigeria

S. Abdulrahman
Department of Agricultural Economics and Rural sociology, Faculty of Agriculture, Ahmadu Bello University, Zaria, Kaduna State, Nigeria

O. Yusuf
Department of Agricultural Economics and Rural sociology, Faculty of Agriculture,
Ahmadu Bello University, Zaria, Kaduna State, Nigeria

A. D. Suleiman
Department of Agricultural Education, Federal College of Education Okene, Kogi State, Nigeria

J. Oloche
Department of Animal Production, University of Agriculture, Makurdi, Nigeria

O. I. A. Oluremi
Department of Animal Nutrition, University of Agriculture, Makurdi, Nigeria

J. A. Paul
Department of Animal Production, University of Agriculture, Makurdi, Nigeria

R. R. Silva
Graduate Program in Animal Science, State University of Southwest Bahia (UESB), Itapetinga-BA, Brazil

A. C. Oliveira
Department of Animal Science. Federal University of Alagoas (UFAL). Arapiraca Campus.
AL Alagoas, Brazil

G. G. P. Carvalho
Department of Animal Science. Federal University of Bahia (UFBA) Campus Salvador, Bahia, Brazil

F. F. Silva
Department of Animal Science, State University of Southwest Bahia (UESB), Itapetinga-BA, Brazil

F. B. L. Mendes
Department of Animal Science, State University of Southwest Bahia (UESB), Itapetinga-BA, Brazil

V. V. S. de. Almeida
Department of Animal Science. Federal University of Alagoas (UFAL). Arapiraca Campus.
AL Alagoas, Brazil

L. B. O. Rodrigues
Graduate Program in Food Engineering, State University of Southwest Bahia (UESB), Itapetinga-BA, Brazil

A. A. Pinheiro
Goiás Agency for Technical Assistance and Rural Extension Agricultural Research (EMATER),
Goiânia-GO Brazil

A. P. G. Silva
Department of Animal Science. Federal University of Bahia (UFBA) Campus Salvador, Bahia, Brazil

R. M. do Prado
Graduate Program in Animal Science, State University of Maringá (UEM), Maringá-PR, Brazil

R. R. Silva
Graduate Program in Animal Science, Southwest Bahia State University, Brazil

A. C. Oliveira
Graduate Program in Animal Science, Federal University of Alagoas, Brazil

G. G. P. Carvalho
Graduate Program in Animal Science, Southwest Bahia State University, Brazil

F. F. Da Silva
Graduate Program in Animal Science, Southwest Bahia State University, Brazil

F. B. L. Mendes
Graduate Program in Animal Science, Southwest Bahia State University, Brazil

V. V. S. De Almeida
Graduate Program in Animal Science, Federal University of Alagoas, Brazil

L. B. O. Rodrigues
Graduate Program in Animal Science, Southwest Bahia State University, Brazil

A. A. Pinheiro
Graduate Program in Animal Science, Federal University of Goias, Brazil

A. P. G. Silva
Graduate Program in Animal Science, Southwest Bahia State University, Brazil

J. W. D Silva
Graduate Program in Animal Science, Southwest Bahia State University, Brazil

M. M. Lisboa
Graduate Program in Animal Science, Southwest Bahia State University, Brazil

Nadeem Abbas Shah
Department of Agricultural Sciences, Allama Iqbal Open University, Islamabad, Pakistan

Nowshad Khan
Department of Agricultural Sciences, Allama Iqbal Open University, Islamabad, Pakistan

Raees Abbas
Institute of Agriculture, Extension and Rural Development, University of Agriculture, Faisalabad 4000, Pakistan

Muhammad Hammad Raza
Institute of Agriculture, Extension and Rural Development, University of Agriculture, Faisalabad 4000, Pakistan

Babar Shahbaz
Institute of Agriculture, Extension and Rural Development, University of Agriculture, Faisalabad 4000, Pakistan

Badar Naseem Siddiqui
Pir Meher Ali Shah Arid Agriculture University, Rawalpindi, Pakistan

Farhat Ullah Khan
Department of Agricultural Sciences, Allama Iqbal Open University, Islamabad, Pakistan

Shafique Qadir Memon
Department of Agricultural Sciences, Allama Iqbal Open University, Islamabad, Pakistan

O. Yusuf
Department of Agricultural Economics and Rural Sociology, Institute for Agricultural Research, Ahmadu Bello University, Zaria, Nigeria

J. O. Olukosi
Department of Agricultural Economics and Rural Sociology, Institute for Agricultural Research, Ahmadu Bello University, Zaria, Nigeria

Dickson Stuart Tayebwa
Central Diagnostic Laboratory, Department of Veterinary Pharmacy Clinics and Comparative Medicine, College of Veterinary Medicine Animal Resources and Biosecurity, Makerere University, Kampala, Uganda

Godfrey Bigirwa
Central Diagnostic Laboratory, Department of Veterinary Pharmacy Clinics and Comparative Medicine, College of Veterinary Medicine Animal Resources and Biosecurity, Makerere University, Kampala, Uganda

Joseph Byaruhanga
Central Diagnostic Laboratory, Department of Veterinary Pharmacy Clinics and Comparative Medicine, College of Veterinary Medicine Animal Resources and Biosecurity, Makerere University, Kampala, Uganda

Keneth Iceland Kasozi
Central Diagnostic Laboratory, Department of Veterinary Pharmacy Clinics and Comparative Medicine, College of Veterinary Medicine Animal Resources and Biosecurity, Makerere University, Kampala, Uganda
Kiboga Regional Veterinary Laboratory, Department of Production and Marketing, Kiboga Local Government, Kiboga, Uganda

www.ingramcontent.com/pod-product-compliance
Lightning Source LLC
Chambersburg PA
CBHW050450200326
41458CB00014B/5128